Great Discoveries and Inventions That Have Changed the World

From the End of the 19th Century to the Present

WHITE STAR PUBLISHERS

Project Editor
VALERIA MANFERTO DE FABIANIS
LAURA ACCOMAZZO

Graphic Design
PAOLA PIACCO

Editorial Staff
CHIARA SCHIAVANO, GIULIA GATTI (ICEIGEO, Milan)

Collaborators: CARLO BATÀ, ALBERTO BELLANI, FRANCESCO D'ANTONI, GIULIA GATTI,
MARGHERITA GIACOSA, SIMONE GRAMEGNA, LORENZO MARSILI, ALESSANDRA MASTROLEO,
PRIMO MIRACCA, PAOLO PACI, PAOLA PAUDICE, ELENA ROSSI, LORENZO SAGRIPANTI,
CHIARA SCHIAVANO

Great Discoveries and Inventions That Have Changed the World

From the End of the 19th Century
to the Present

Edited by
GIANNI MORELLI

Contents

Introduction

by Gianni Morelli

Everything around was illuminated by an electric current flowing through thin tungsten filaments in a vacuum enclosed in a glass bulb: people dining, reading, chatting, shows and concerts, roads, bridges, factories, quays, airport runways, stations, and celebrations. Dusk took on another meaning: it no longer indicated the end of the day, but the shift from afternoon to evening. The incandescent light bulb was one of the symbols of the "Great Change". Another one, for example, was the refrigerator, the only electrical appliance that cannot be replaced by human labor. We can boil broth, we can wash our clothes in the river, and we can use a broom instead of a vacuum cleaner. We cannot produce "cold", let alone ice. In fact: for thousands of years, "cold" was carried down from the mountains and preserved in large insulated and underground vaults, which were filled with snow in the winter and then opened in the summer, only once. It was all very costly, and thus only for the very wealthy.

In contrast, now two hundred million fridges are produced in the world every year! The refrigerator, like the light bulb, was an illustrious protagonist of the Great Change, whose real dimensions in terms of quantity and quality we find difficult to realize, because our temporal and cultural references are limited to the horizon of the human life. But what happened in the twentieth century was not a mere step in the evolution of human civilization; it was a momentous break—a leap we can compare to the invention of the wheel, the spread of agriculture, the domestication of animals, and the Industrial Revolution. The nineteenth century formed the theoretical foundations and technical conditions which then enabled the twentieth century to transform not only human life, but also that of the entire planet (and its surroundings).

In one century, we moved from traveling around the world in eighty days to airships, to jet airplanes, and finally to satellites that complete an orbit in one hour and forty minutes, 500 miles above the surface of the oceans. The visit to the Moon is already half a century "old". We progressed from postmen on muleback to radio, and today it is the Web that reduces, or rather eliminates, all distances. We moved from bloodletting to antibiotics to pacemakers to laser scalpels; from fragile and anonymous everyday objects to tough, colored objects made of plastic; from payments in coin to payments by credit card; from the quill pen to electronic mail and voice messages; from the telegraph to fiber optics, that spider's web that now extends underneath all the cities of the world. Everything is fast, instantaneous.

Time is the most striking variable in this process: the acceleration that began after the middle of the nineteenth century was a shock. Generations began keeping pace with each other and pursuing the extraordinary changes proposed and imposed by visionaries, scientists, inventors, businessmen, and states people. Perhaps it interrupted forever that genealogical continuity where the old were the only repositories of a universal wisdom, independently of the passing decades, of chronology, of epidemics, of wars. The digital world is now creating an ever-greater divide between children and parents. The latter are condemned to follow, with immense difficulty, the dizzying evolution of technological progress. Today the smartphone is in everyone's pocket or purse, but it is often, perhaps excessively, in everyone's hand. It is the direct descendant of simple portable telephones, the so-called cell phones, and it has rapidly become the most effective bridge to greater integration between the technology of communication and the human body, between microprocessors and the nervous system. As we await the completion of this process, we are also designing sophisticated robots and artificial intelligence that are capable of learning from experience and of learning independently.

In the past, there had been transformations, almost equally dramatic, from photography, the cinema, television, the gramophone, and radio, which completely changed our perception of the world and space, and which redefined our culture. Let us not forget progress in our understanding

Introduction

of DNA, the human genome, the invisible but inexorable mechanisms that create individuals as unique members of the same species.

In short, everything changed in better part of the last century. Today, space extends from the infinitesimally small to the infinitely great, from neutrons to the unknown and fascinating limits of the universe. Humans push the envelope constantly, and at times impudently, as in the case of genetics, the internal and external limits imposed on them by nature, in an endless struggle full of resounding defeats, but also of exciting and unexpected victories. The passing of time collapses into the present, and in the continuous struggle for survival physical effort is increasingly superfluous. The old world no longer exists, the scenarios for progress are shaped by the technological revolution and by the small, but great, inventions that have transformed our life, improving it and making it more convenient and inevitably more complex.

Some of the countless steps on this extraordinary journey have been longer and more decisive. Here we relate them and celebrate their protagonists: the men and women who imagined and translated intuitions and discoveries into real or virtual machines, who brought them into everyday life and culture, transforming them forever.

Certainly, not all the news is positive: for example, we'll need between five hundred and a thousand years before the huge islands of plastic that have formed in the middle of the oceans degrade. And there is more: today, more than five hundred nuclear power stations are working in about forty countries, on all the continents except Oceania. Every day, these power stations produce nuclear waste that will remain radioactive for hundreds of thousands of years. Who knows whether it will outlive us, or if we'll manage to survive it. And the latter will mean that other progress that is unthinkable today will have enabled us to do so.

What seemed to be science fiction now belongs to our past: projects, grandiose dreams, eternal hopes, so much hard work, and an inevitable stroke of luck, just the right amount. Of the many changes, one is perhaps more profound than the others. Today we are ready for everything, and nothing really surprises us any more. We read and talk normally about the forthcoming voyage to Mars, and about the construction of probes to send toward Alpha Centauri, the closest star: 4.3 light years, twenty-five trillion miles from here. Traveling at the planned speed of 37,000 miles per second, they will take twenty years to reach their destination. It is true that we are talking about microscopic probes, but it seems that now, despite the present limits, a voyage in the galaxy is only a question of time, and not even such a long time if developments continue at their present rate.

In the same way (apparently), another aspect of physics, quantum entanglement—combined with the uncertainty principle—seems to suggest that everything is a product of probability, including ourselves; it acquires reality only when the particles of the microscopic world come into contact or are measured. And if two particles do come into contact, they become Siamese twins: what one does, the other will do, even if it is light years away. To summarize the scientific deductions in simplistic terms, it means that nothing is impossible, but only highly improbable. It is better not to explain this more thoroughly, because it would take too long and would probably be too disturbing.

Thus, perhaps, there will not be another time of marvels. At least, not like the men and women who listened, transported and frightened by the sound coming from the phonograph horn, wondering where the trick was, where the singer and pianist were hiding. Perhaps they hoped that there really was a trick so as not to think of magic, of the devil or (who knows?) some other witchcraft. Because in the end, both then and now, it is easier to believe what we see *a priori* than ask questions that are too complex and demanding. However, sooner or later, we'll really fly high, where even imagination struggles to comprehend the geography and the principles of the universe. And then still higher up; who knows? Perhaps we will open the doors of parallel universes.

5 An array of optic fibers.

10-11 August 13, 2005, NASA STS-114 mission: the astronaut Stephen K. Robinson, tethered to an arm of the space station, is engaged in an extra-vehicular activity (EVA) session.

TECHNOLOGY

Combine Harvester

Alternating Current   Plastic

Assembly Line

Optic Fiber   Nuclear Energy

Solar Panels

Humans and Space

Laser   Robot

What is the common feature of the combine harvester and the laser, the assembly line and satellites, nuclear energy and plastic? They are solutions to practical problems, the optimizations of procedures, the results of experimentation and of scientific research. Behind them lie mechanics, electronics, ideas, and organization, but also strokes of luck. In short: technology.

After all, what would an automobile cost without the assembly line and without plastic? What type of electrical current would we have used for our stereos if Nikola Tesla had not been hired by George Westinghouse? What materials would transoceanic telecommunications cables have been made of if it had not been a sunny day in Geneva when Jean-Daniel Colladon discovered the phenomenon underlying the use of optic fiber? And when would we have seen the blue planet on the black background of space without satellites or space capsules?

There is no answer to these questions; but it is clear that, thanks to all this, not only human life has changed, but also the entire planet, for better or worse. We can now build machines in our own image, illuminate the Earth, harvest all the wheat on a great plain in only a few days, cover enormous distances in a short time, exploit invisible particles of matter to produce energy, power a refrigerator with sunlight, and synthesize new materials to create virtually indestructible objects.

Certainly, we know that technology not only produces wonders, but also side-effects, sometimes very dangerous ones, like radioactive waste or the islands of plastic in the oceans. But it is difficult to imagine that the mechanism can be stopped. Perhaps the future already lies in robots that will clean everything up, perhaps to start the game again from the beginning.

Combine Harvester

At the end of the nineteenth century, the first agricultural machines supplemented the work of humans and animals. With the first McCormick harvesters—in the photo we see a horse-drawn model—agriculture entered a new era that would transform it completely.

The great revolution in agriculture

Certain inventions create a break in the progress of time, overwhelm society, concentrate it, and reorganize it in a completely new structure. Today we see these inventions without understanding their importance, but in a short time they have led to the "world as we knew it" as a distant, faded memory. The combine harvester is an example.

The Ancient Romans already used a rudimentary mechanical harvester, which was then "forgotten" in the following centuries. But in 1831 the American Cyrus Hall McCormick designed and built a new mechanical harvester that cut stalks and gathered heads, and which became the forerunner of modern combines. At the beginning of the nineteenth century, and until the middle of the twentieth, the economy and society were still principally founded on agriculture. Throughout the world, the fields were filled with multitudes of peasants intent on cutting, threshing, gathering, and working every kind of cereal crop. From McCormick onward, their work was gradually brought together in a single machine: the combine harvester. The combine was first drawn by animals and then driven by steam engines, but today it is powered by biofuel. This extraordinary machine constitutes the center of a hitherto unseen growth in agricultural production. This process has been favored in recent decades by fertilizers, scientific innovation, research, and ever more futuristic methods in agriculture.

Actually, in the early stages the combine was both innovative and controversial. While in the United States its introduction made it possible to make up for the lack of manpower, as it was well suited to the limitless flatlands to cultivate, in Europe it spread more slowly because of the high costs and farmhands' opposition: they saw it as a threat to their employment. Very quickly, their fears were proven well founded and farm workers were forced to say farewell to a world that had been eternally linked to the sweat of the brow and cuts on the hands.

The countryside was depopulated, the cities began to expand, and new occupations and opportunities were created. Social relationships changed as well. At the same time, the curve of the economy drawn on the Cartesian plane rapidly assumed a new form, changing agriculture and with it the world. While previously most of the population was employed in the fields, in less than two centuries the percentage of workers engaged in agriculture fell to under 5%. The extraordinary contrasting phenomenon is soaring agricultural production, allowing it to meet today's apparently relentless increase in the world's population.

16 Designed by the Briton Dan Albone, the Ivel Agricultural Motor was the first mass-produced tractor with an internal combustion engine. It was light and efficient, and in 1903 cost the moderate sum of 300 pounds.

16-17 1954, Haute Provence, France. In large fields of crops, farm workers sit comfortably at the controls and drive a noisy, futuristic engine-powered Harris "agricultural spaceship".

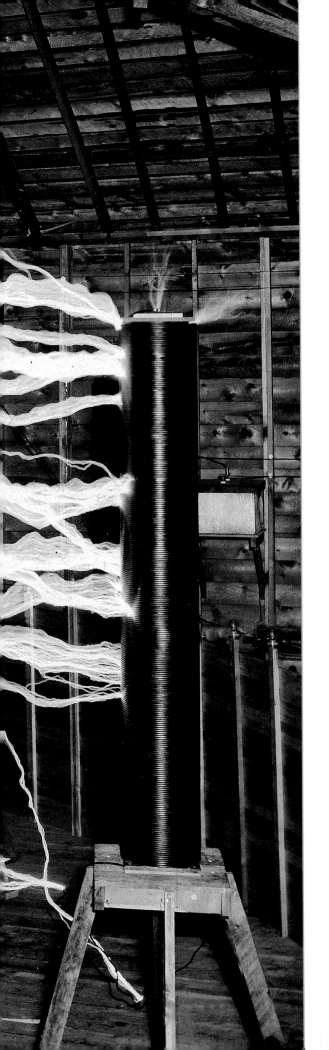

Alternating Current

Nikola Tesla, the man who electrified the world

AC/DC: these initials have left their mark on history, not only on rock music. The so-called current war was a battle between the two greatest inventors of the nineteenth century, Thomas Alva Edison and Nikola Tesla. The stakes were high: electric power transmission over long distances to light up the streets and the lives of millions of people.

Both of them were telegraphers and brilliant scientists. They followed different routes, or rather, different currents. Edison saw science as a means of profit, whereas Tesla aimed to improve human life.

On September 4th, 1882 Pearl Street Station, the first commercial power station in the world, began to generate direct current in the south of Manhattan, lighting with 400 lamps the homes and stores of 85 residents, customers of the Edison Illuminating Company. It was a success! However, direct current, a flow of electrical charges passing always in the same direction at a constant intensity, was too costly, and was subject to serious dispersion problems over long distances.

Nikola Tesla, the "wizard" of electricity, in the Pikes Peak laboratory in Colorado, in 1900 ca. His high-voltage, high-frequency transformers discharged into the ground quantities of energy greater than those of lightning.

Edison was far-seeing. He was that twelve-year-old newspaper boy who had rapidly succeeded in opening his "patent factory", turning out the electric light bulb, the phonograph and the kinetoscope, but he did not perceive the potential of the "rising star", alternating current. As a matter of speaking, Edison was "deaf in that ear" not only because of an actual physical problem, which began when as a telegraphic operator his employer found him pouring sulfuric acid under the desk during one of his chemical experiments, and hit him in a fit of rage, causing permanent damage to his hearing.

But Tesla's hearing was excellent, and he had understood the potential of alternating current. Constituted by a flow that changes direction and intensity at regular intervals, following a sine wave, today it is the most widespread and supplies all the electrical systems in our homes.

Tesla was a Serbian engineer who emigrated to the United States in 1884. From an early age, he had daydreams about dazzling flashes of lightning and tongues of fire. They were not only fantasies: in fact, he put on a show before a dumfounded and ecstatic audience in his laboratory on South Fifth Avenue in New York around 1891.

Alternating Current

Tesla had no need to prepare models and prototypes—he designed his inventions completely in his head. He shocked everyone, colleagues and workers, by his unusual approach. Everyone except Edison, who hired him but then deceived him. In order to warn American housewives and at the same time discredit Tesla, in 1888 Edison even invented the electric chair to show the dangers of AC due to the higher voltage it developed compared to DC. To refute his rival, the Sorcerer of Lightning allowed a low-voltage AC current to pass through all his body in a voluntary electric shock treatment. In 1886, the American entrepreneur George Westinghouse founded a company that aimed to build alternative electric power stations to those of Edison. The magnate bought all the patents on AC motors and brought Nikola Tesla under his wing. Westinghouse Electric was destined to win the market and mark the Second Industrial Revolution.

The first AC induction motor—the principle of which had been discovered by Michael Faraday in 1831—was built by the Serbian engineer Nikola Tesla. In 1882, Tesla was hired by the Continental Edison Company, where he advanced the study of alternating current.

Plastic

One name for so many synthetic materials

Plastic is chemistry in the "from test tube to business" version. Synthetic products developed in the laboratory rapidly become consumer products.

Its history consists of names that practically all of us know because they are with us in every place and at every moment of the day. From celluloid, in its various versions, synthesized for the first time in 1861 by Alexander Parkes, to make handles and boxes, to Bakelite—which in 1907 took its name from its creator Leo Baekeland (in which many things can be created, including the extraordinary Art Deco objects of the era). From thin, flexible, transparent, impermeable cellophane sheets (Jacques Edwin Brandenberger) to the mother of all synthetic fibers, Wallace Carothers' nylon (1935) which was formed into clothes, linen items, curtains, and even parachutes. The '30s were also the years of polyethylene, used for making new bottles for water and soft drinks, and Fritz Klatte's PVC (polyvinyl chloride), used for producing 45 and 33 RPM records.

In the meantime, durability, molding techniques, colors, etc. improved. The natural sources were rubber and increasingly oil, and thus the '50s were "plastic years": nylon stockings for ladies, Formica furniture, polyethylene furnishings, and celluloid cinema films were their symbols.

Plastics definitively burst into the world of creativity, art, design, and fashion. Electrical appliances and automobiles contained plastics in ever greater measure.

Today, with technopolymers, unimaginable results have been obtained: materials that are more resistant than steel (to impact, to temperature, to solvents): they are invaluable for hospitals, airplanes, armies, factories, ships, satellites, power stations.

A formaldehyde and phenol resin formed in hot molds: the simple recipe for Leo Baekeland's creation, Bakelite. Its ease of production, strength, and bright colors ensured its success.

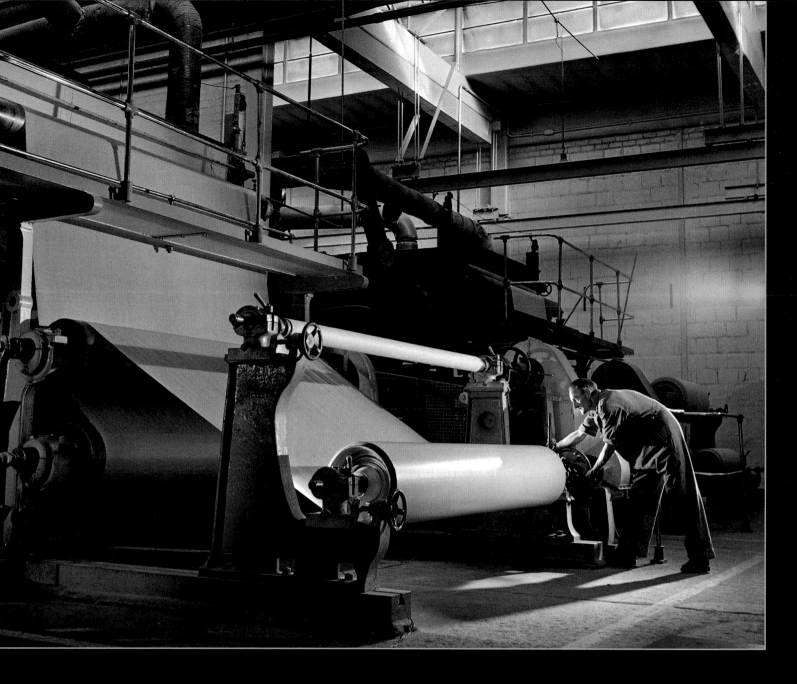

The probes we have sent into the solar system are mostly assembled from plastics. Certainly, some plastic has stayed on the Earth: in the Atlantic and the Pacific, the movement of currents has formed two "artificial islands" as large as Belgium and the Iberian Peninsula. They are made up of large and microscopic plastic objects which have been thrown into the sea or rivers, from plastic shopping bags to larger objects. These are materials that will not "dissolve" for hundreds of years, are lethal for marine fauna, and are non-recyclable. Some have denounced the situation but, as of today, countermeasures do not seem to be enough.

24 Later, producers added to the basic formula catalysts, capable of giving the Bakelite greater plasticity or strength. In the photograph, we see the process of rolling great Bakelite ribbons of varying thicknesses.

25 The Philco Radio & Television Corporation in Perivale (a London suburb), in 1936: the assembly line for the new model called People's Radio. This radio was designed to be mass produced in order to keep the price very low. The appliance was manufactured in Bakelite, the first plastic material used for radio production. It could be molded perfectly and was also an excellent electrical insulator.

Plastic

26 Nylon could be transformed into thread of differing thicknesses and sheen; still today it is considered the best artificial textile fiber.

27 Top left Records made of vinyl (or PVC, polyvinylchloride) appeared in the United States in 1948, while in London they were produced and marketed from 1950 onward.

27 Top right Water/wear-resistance, low flammability, and complete water repellence: these are mechanical properties that have made PVC extremely versatile.

27 Bottom Light, transparent, water resistant, and resistant to micro-organisms: cellophane, a widely-used plastic, from umbrellas to film for food products.

Assembly Line

From Ford plants to robots

Detroit, Michigan, October 7th, 1913: the assembly line was born in the Ford plants. The idea was simple: a conveyor belt on which moved the parts to be assembled. Along this, there was a line of workers who gradually produced the finished product, their timing synchronized in standardized processes: scientifically organized work, drastic reduction in assembly time, increase in productivity by a factor of 8, soaring profits. In a process subdivided into 45 steps, the Model T—the manufacturer's flagship product—was produced in only 93 minutes. And every 3 minutes a new one was ready for an American middle class family to climb aboard. No training courses nor particular abilities were necessary: anyone could be on the assembly line, including women. Wages increased, working hours decreased. Entrepreneurial qualities were masked by philanthropy. There were no tricks nor deception: this was Fordism, even if it had yet to be given that name.

Previously, there had been similar rudimentary attempts in shipyards in Britain, the cradle of the Industrial Revolution, but these had been limited in time and space. Henry Ford had been struck by the organization of the work in the Swift & Company slaughterhouse in Chicago, where the process was actually the reverse: the carcass of an animal (instead of a future automobile

In 1936, after the Wall Street Crash and in the midst of the Great Depression, Charlie Chaplin wrote, directed and starred in the film *Modern Times*: a worker on an assembly, who cannot bear the frenetic pace and repetitiveness of his task, has a nervous breakdown and ends up in a clinic.

frame) moved on a conveyor belt, where it was butchered step after step by the chain of operators.

And what happened to the workers who spent the whole day on the assembly line, repeating the same movement mechanically for hours, manning the line unselfishly and assiduously? The critics said "alienation", but you can't stop progress for such a detail. "Marx is dead. I'm not feeling that good myself," in the words of Eugène Ionesco or

30-31 Assembling the chassis on the frame of automobiles leaving the Ford plants. Detroit, 1913.

31 The American entrepreneur Henry Ford: inventor of the modern automobile industry.

32   Women contributing to the war effort during the Second World War as they check bomber cockpits lined up in the yard of an American aircraft plant.

33   The highly automated assembly line of the car manufacturer Opel in Russelsheim am Main, Germany: today the presence of humans is limited to monitoring and supervision.

Woody Allen (we don't know who said it first)... It's no accident that the most searing critique of the assembly line came precisely from a great comic actor, Charlie Chaplin, who in *Modern Times* had the very undemanding task of frantically tightening bolts on anonymous metal plates.

With the advent of automation, robots have replaced humans above all in the most dangerous and repetitive tasks. If we discount Isaac Asimov and his short stories, and the series of Apocalyptic Hollywood films, humans are left, at least for now, with the design and its realization, as well as the programming of the machines, which then work independently.

Optic Fiber

The fastest communication is through glass

University of Geneva, 1841. In a dark lecture room, a young physics professor, Jean-Daniel Colladon, channeled the light from the window with a reflecting metal funnel to illuminate the desk where he was showing the audience a hydraulics experiment. When the light accidentally struck the hole in a container from which there was a jet of water, the spectators were astounded: the light followed the fall of the water as if it were trapped in it. By chance, that young professor had discovered the phenomenon of total internal reflection, the fundamental principle underlying optic fiber. In the meantime, it was used for theatrical sets and the design of fountains. In the staging of *Faust* at the Paris Opéra, Mephistopheles causes a bright red stream to come out of a barrel: it was a water jet, passed through a red glass tube illuminated by Colladon's trick.

It was the creator of polarized sunglasses, Clarence W. Hansell, who in 1926 intuited the potential of the total internal reflection phenomenon and proposed glass fiber as the most suitable means for the transmission of images. A restless and impatient researcher, the holder of over 300 patents, Hansell soon lost interest in the project, but experimentation went on. Fiber covering enabled Basil Hirschowitz' team at the University of Michigan, in 1957, to achieve the first glass fiber gastroscope, which Basil tested on himself.

Finally, in 1965, Charles K. Kao discovered that glass fiber can replace copper wire as an efficient means of telecommunication, provided some minimum requisites are met to guarantee the quality of the transmission. Today all the main telephone and Internet links use optic fiber, and undersea cables connect the various continents. The longest at the moment in use is the Asia-America Gateway (AAG), which links California and Singapore—it is more than 12,000 miles long.

Around 1960, Walter P. Siegmund, a researcher and scientist at the American Optical Corporation, a Massachusetts optical products company, demonstrates the potential of fiber optic cables.

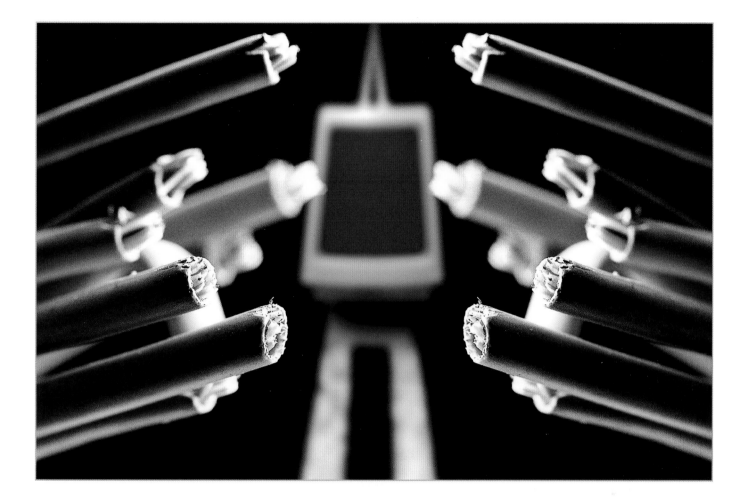

The speed of optic fiber seems to arrive everywhere. Like the mythical steed of a Hindu legend, which inspired a French artist: he symbolically covered a horse in fiber optic material. Yes, because this is the magic of fiber: this versatile material can create fantastic artistic installations, help doctors in endoscopies and precise surgical interventions, and permit the ever vaster, faster, and safer spread of information.

36 e 37 The Integration Laboratory of Telstra, an Australian telecommunications company. In 2015, it collaborated with Ericsson Telstra to achieve a fiber-optic link—at 100 gigabits per second—over a distance of more than 6,2 miles between the cities of Melbourne, Sydney, and Perth. On the right, sections of fiber-optic cables: they offer the highest transmission speed possible.

38-39 Detail of the sculpture *Seed Cathedral*, whose construction was coordinated by the British architect Thomas Heatherwick. Exhibited at the Shanghai Expo in China in 2010, it was over 65 feet high and composed of 60,000 transparent optic fibers containing plant seeds.

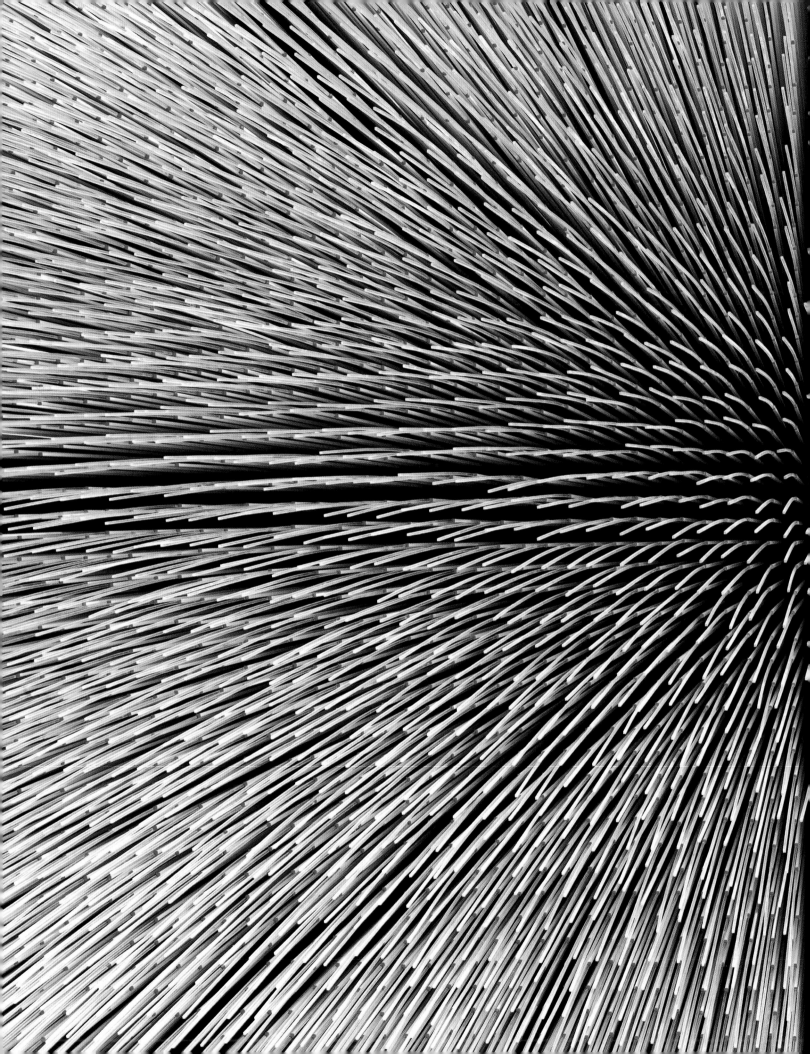

Nuclear Energy

Nuclear fission, the atomic bomb, nuclear power stations

To rewrite the basic principles of the composition of matter: this was the goal of the extraordinary theoretical and experimental effort carried out by physicists and chemists across the world, among whom 16 future Nobel Prize winners. It had extremely important consequences in the political, military, and social fields in the first decades of the twentieth century.

From Becquerel's experiments (1896), to Einstein's ideas (the famous law of the equivalence between mass and energy, 1905), to Fermi's applications (the atomic battery, 1942), hypotheses chased and succeeded each other, with formulations, experiments (and some arguments), in an unprecedented scientific competition. An intellectual challenge that started modestly and forced scientists to rewrite theories of the formation and development of the stars and the Universe itself.

A difficult matter, populated by elusive protagonists, which sometimes behave like corpuscles and sometimes like waves, difficult or impossible to observe without disturbing their behavior (Heisenberg's Uncertainty Principle, 1927).

In chronological order: in 1932, Chadwick discovered the neutron; in 1934, Fermi and the "Via Panisperna boys" (*I ragazzi di via Panisperna*, a group of young physicists at the University of Rome comprising Amaldi, D'Agostino, Majorana, Pontecorvo, Rasetti, and Segrè) used neutrons to strike systematically the nuclei of all the atoms of the periodic table. Fermi discovered that neutrons slowed by a layer of paraffin interacted more effectively with the nuclei of the target elements. In 1938, Hahn and Strassmann, by repeating Fermi's experiments, noted that in this process lighter atoms were generated, and not heavier elements, as had been thought until that time.

In 1939, Meitner and Frisch laid the theoretical bases of fission: by applying Einstein's law they "explained" the relation between mass and energy observed in the experiment; the uranium nucleus is unstable, and after capturing a neutron it can divide into two parts.

After receiving the Nobel Prize for Physics at only 37 years, Enrico Fermi, whose wife Laura Capon was Jewish, openly dissented with the Fascist regime. In December, 1938, he sailed for the United States. We see him here in 1942, in the control room of Chicago Pile-1, the first nuclear reactor in the world.

The research took place to the incredulity and astonishment of the scientific community, and of the protagonists themselves. Everything led towards unexpected results. It seemed that from one element others were generated different from the first, a process resembling more an alchemist's cauldron than a scientific laboratory.

But in the '40s, scientific competition moved into the field of war, with each of the two sides fearing that their adversary could succeed first in producing weapons using the enormous

42 On the left, General Leslie Groves, military chief, and Robert Oppenheimer, scientific director of the Manhattan Project; during the Second World War, the project led to the development of the first atomic bombs.

42-43 The Calutron, the acronym of California University Cyclotron, was a mass spectrograph used to separate fissile uranium-235 from uranium-238, as part of the Manhattan Project. Only decades later did those involved learn of the purposes of their work.

energy liberated by the fusion process. In this context, the Manhattan Project was born: Gadget, Little Boy, and Fat Man were nicknames for lethal weapons, produced through a commitment which began as collaborative international research, but that concluded on the barricades of ideological, political, and cultural differences amid the ruins of war.

"It is harder to crack prejudice than an atom." (A. Einstein)

Since the end of the war, nations have begun to reconsider the use of fission for peaceful purposes, with the construction of nuclear power stations. In the '90s, they reached over 17% of the nuclear power produced in the world; this gradually decreased to 10.5% in 2016.

44 July, 1974: Vice President Gerald Ford meets the scientists of the Scyllac program, which was part of research on thermonuclear interactions and the properties of plasma, carried out in those years in the laboratories of Los Alamos in New Mexico.

45 Technicians at work in a nuclear fission power station. The first power stations of this type for the production of electrical energy were commissioned halfway through the twentieth century in the Soviet Union and in Britain.

46 A doctor examines images obtained with magnetic resonance imaging (MRI). With this method the patient is not subjected to potentially harmful radiation, as happens with other medical investigations, for example CT.

47 An image of the brain obtained with computerized tomography (CT), a diagnostic imaging technique that exploits ionizing radiation (X-rays) to scan sections (tomography) of the human body.

Nuclear Energy

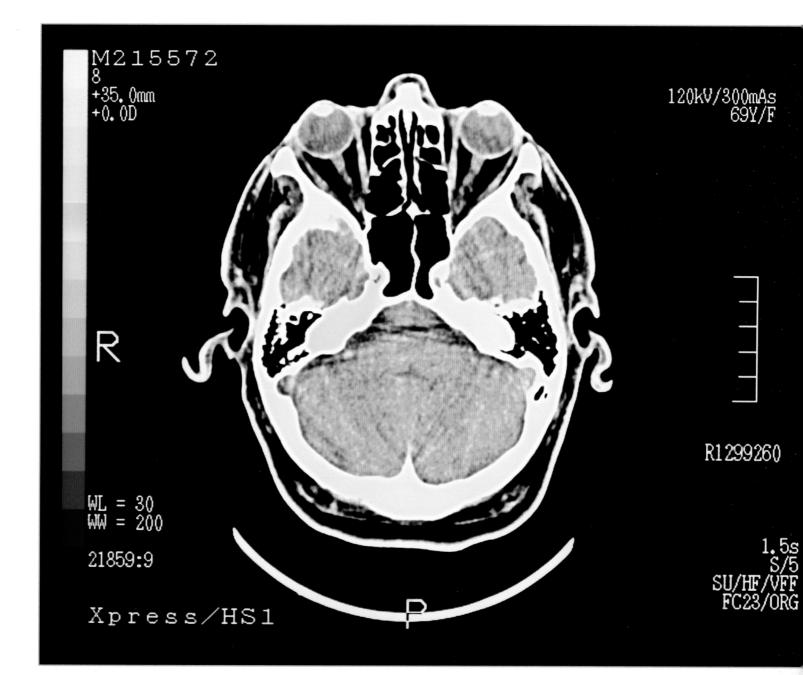

Very recently, nuclear physics has given rise to an entire branch of medicine, nuclear medicine, and a new class of drugs, radiopharmaceuticals. The use of radiodiagnosis and radiotherapy today is widespread and irreplaceable, in particular in treating tumors. Among the most widespread applications, computerized tomography (CAT, today CT), positron emission tomography (PET), and nuclear magnetic resonance (NMR).

While nuclear waste from hospitals is low/medium activity, with a treatment cycle of a few hundred years, the real problem is represented by waste from power stations, which is active for hundreds of thousands of years. Will someone, if there is anyone, be able to explain this awkward presence to our very distant descendants?

Solar Panels

Green energy for a clean world

Large transparent surfaces in blueberry, strawberry, and melon colors: these are the windows of a large veranda, but they are also a small power station—the most recent and least polluting—which feeds power to the entire home.

For some billions of years, the Sun has supplied energy to the Earth, but only in the last few decades have we begun to consider whether it is possible to derive energy from its light. To do this, devices such as solar cells are necessary, which exploit the capacity of some materials to produce electricity by simple exposure to the light: at the moment, the most widely used is monocrystalline silicon.

The first cell was made from silicon in 1953, but as early as 1957 the yield (the percentage of solar energy transformed into electrical energy) rose to 15%, which is comparable to the yield of most cells used today. The production of solar energy is increasingly economical and versatile in its applications and is constantly growing, as it has more than quintupled in the last five years.

It is already possible to build the roof of a house in photovoltaic tiles, very similar to the traditional ones in earthenware or slate, as well as photovoltaic windows, low walls and window panes, and even photovoltaic road surfaces. New transparent materials made of graphene enable us to design "active façades" in which the different components (windows, curtains, shutters, and front panels for balconies) are all able to produce energy: the so-called integrated photovoltaic system.

Specialists are also studying solar panels integrated in long, extremely thin, flexible optical fibers, which can be woven together with other materials. One day clothes, too, could be capable of supplying energy for common devices or health monitoring.

A vast area of solar panels, feeding a photovoltaic plant, with a service road running through it. The Sun supplies the Earth with a quantity of energy thousands of times greater than the planet's energy needs, but it still little exploited.

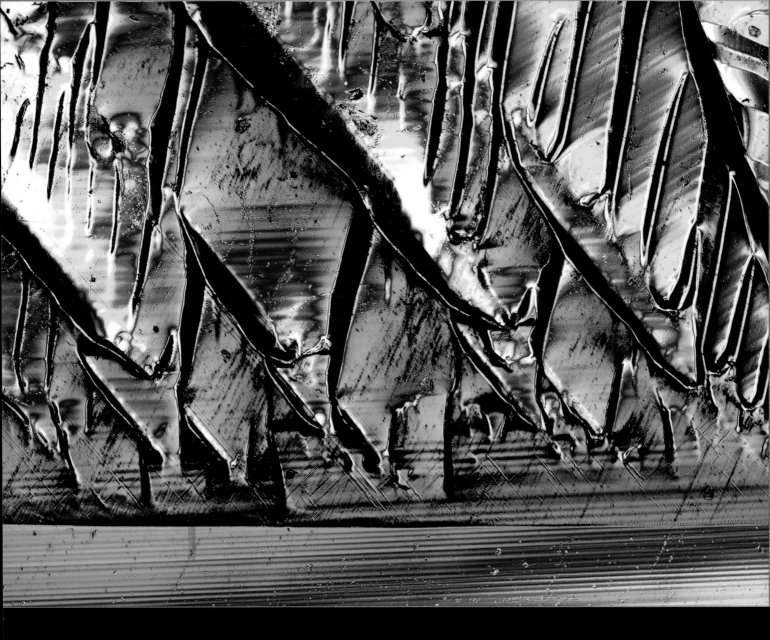

The latest and most recent alternative is Grätzel cells, from the name of the German chemist who created them: these are particular photoelectrochemical cells that use glass and organic molecules extracted from fruit to partially reproduce the photosynthetic process. They promise high efficiency with little solar radiation, less use of expensive materials, and lower production costs.

In the meantime, we send satellites to explore the solar system at a distance of thousands of millions of miles that send back information and photographs after years of travelling, only thanks to solar cells that make it possible to power the on-board instruments for such a long time.

50 Microphotograph of a polycrystalline silicon sample, about 0.16 inches in size, used for the construction of photovoltaic panels. The material needs to be 99.999999% pure.

51 Students survey Solar panels mounted on a Shuttle at the Space Camp in Huntsville, Alabama. Exploiting solar energy 24 hours a day is also the objective of a project by the Japanese space agency, which is working to construct a giant photovoltaic plant that will orbit 22,4 miles above the Earth.

Humans and Space

A Home beyond the Earth

Two things have played a crucial role in the history of space exploration: power and propaganda. The technologies that can be used for peaceful and for military purposes are called "dual-use", and space technology is one of the most important of these.

The first step: rockets capable of reaching a sufficient altitude to enter the Earth's orbit, the so-called Karman line (60 miles above sea level), where it is considered that "space" begins. The first real rockets were the V-2s used by the Nazis during the Second World War to bomb various European cities; these were then inherited by the Americans, together with the engineers, also Nazis, led by Werner Von Braun. In fact, on June 20th, 1944, a V-2 was the first human artifact to reach the Karman line. After the war ended, the V-2s provided the first image of the Earth from space, (a 35mm black-and-white photograph) in October, 1946.

The so called "space race", which broke out during the Cold War between the USSR and the USA, transformed science fiction into reality. Sputnik 1 became the first artificial satellite in orbit (October 4th, 1957). Laika was the first animal

The International Space Station (ISS) in orbit, viewed from the space shuttle Atlantis as it moves away after separation. It has just completed Mission STS-122 by transporting new instrumentation aboard the ISS and bringing back instruments that were no longer operative.

to orbit the Earth, on Sputnik 2. Yuri Gagarin was the first human being hundreds of miles away from his planet, on Vostok 1, on April 12th, 1961. Then came the probes toward the Moon, Venus, and Mars. And finally Neil Armstrong walked on the Moon on July 20th, 1969, with Apollo 11.

Since then, space exploration has changed scale. Astronauts have worked on space stations, in international crews. By now, Saliut, Skylab, Mir, and ISS are familiar names and initials. Astronauts have worked on great telescopes like Hubble (in orbit since 1990), which have "seen" the universe in a way that would never have been possible from the Earth.

54 and 54-55 Werner von Braun (far right) was a pioneer in missile engineering for the Third Reich. Following the Second World War, he crossed over to work for the government of the United States. In the photo on the left, we can see a V2 ready for launch from the White Sands range, New Mexico, USA, in 1946.

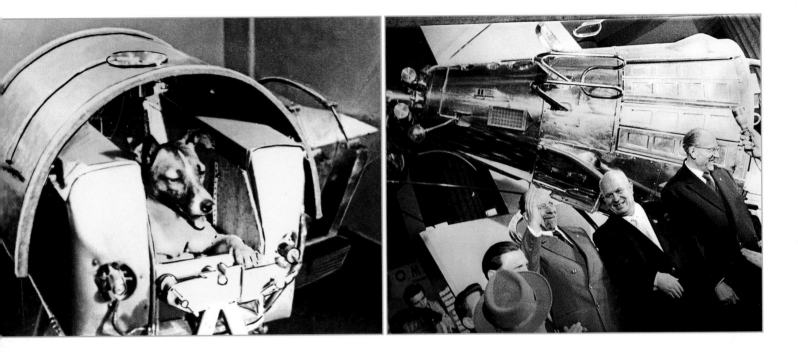

Nations have worked on probes to send toward other heavenly bodies. Already in the '70s, Viking and Venera had reached Mars and Venus. And they had continued to work on the great problem: keeping vehicle and instruments "alive" for the several years necessary to reach and study the solar system.

We have received, and in some cases still receive, extraordinary images and precious scientific information from Messenger (Mercury), Rosetta (Comet 67P), Juno (Jupiter), Cassini (Saturn and its rings), New Horizon (Pluto), and so on. The difficulties are the distances: Voyager 2 flew over Neptune in 1989 after travelling for 12 years. The energy supply and the communications are key: the signal becomes weaker and weaker because of the distance (for example, it takes more than 4 hours from Neptune to reach the Earth).

56 Little Laika (in the photo on the left), the first mammal to orbit the Earth, on Sputnik 2 (November 3rd, 1957). The head of state of the Soviet Union, Nikita Khrushchev (the second from the right in the photo on the right), poses together with other leaders in front of Sputnik 3 at the Leipzig Fair in March, 1959.

57 The legendary smile of Yuri Gagarin, the first man in Space, who orbited the Earth on the Soviet spaceship Vostok 1 on April 12th, 1961.

Today nations are working on landers: robots roaming freely over the surfaces of planets and satellites. People are also working to take humans to Mars, which means conceiving of producing enough oxygen and food for at least 500 days, for a crew capable of staying a month on the Red Planet. Everything is envisioned for 2025-2035.

According to astronomers' estimates, in our galaxy alone there are 100 billion stars and probably double that number of planets. Therefore—excluding satellites (in the sense of heavenly bodies), asteroids, comets, black holes and so on—there are at least 3 billion temptations for a people eager for knowledge and opportunity like (the best of) the human race. Sure, the distances are formidable, but who knows? . . .

58 The American astronaut Buzz Aldrin walks in the Sea of Tranquility. He was the second man to touch the Moon's surface, on July 21st, 1969.

59 The unforgettable image of the Blue Planet rising above the horizon of the Lunar Module, which is taking astronauts Armstrong and Aldrin back to the Apollo 11 Command Module (commanded by Michael Collins) after their long "walk" on the surface of Earth's satellite.

60   The MIR Space Station orbited the Earth 86.331 times between 1986 and 2001: in 15 years, it covered more than 2,2 billion miles at about 250 miles above the Earth's surface. It hosted scores of astronauts, men and women, and enabled scientists to perform many analyses and to gather information on Space and the atmosphere. We see it together with the shuttle Atlantis.

61   The Phoenix Mars Lander at work on the surface of Mars. Launched on August 4th, 2007, Phoenix touched the surface of the northern polar region on May 25th, 2008. The lander examined samples of Martian soil in search of water (which it found) and possible traces of bacteria. It sent information for almost six months (three more than planned), and then there was silence. In 2010, further contact was attempted, unsuccessfully, and the mission was formally ended.

Laser

The ray of light that has revolutionized medical techniques, transmissions, and space measurements

With the light of a laser one can identify the interaction between a certain electron and its nucleus. Or else one can measure the distance between the Earth and the Moon with incredible precision. The light of a laser is magical.

Since 1960, when it was first made (T.H. Maiman), the laser has rapidly become the most widespread high technology device, from the purely scientific context (including chemical-physical analysis) to the contexts of space, information technology, industry, medicine and art, and even audio-video and entertainment. An acronym of "light amplification by stimulated emission of radiation", the laser produces bands of light characterized by considerable directionality, like the famous Jedi swords of the cinema. The ray of light is generated by an active medium—a crystal, a gas or a solution—in a cavity of which the two extremities are mirrored surfaces, one of them semi-reflective. The active medium, "excited" electromagnetically, produces photons that rebound from the walls of the photoconductor: only those that are perfectly aligned ("collimated") escape through the semi-reflecting mirror in the form of an extremely narrow and strongly monochromatic ray (formed of light of the same frequency, thus of the same color).

Bar code readers, found in every sales outlet, are lasers; while CD and DVD players, which are also based on the laser, seem to be in decline.

In optical fibers, the laser is made to move inside a glass filament, making it possible to transmit data at high speed.

In medicine, they are used in the surgical treatment of myopia, in lithotripters for the treatment of gall or kidney stones, and in high-precision scalpels.

Theodore H. Maiman at the Hughes Research Laboratories in Malibu, California, on May 16th, 1960. The American engineer activates the first laser light by using a ruby crystal in the shape of a cube as an active medium: the crystal absorbs the light and projects a beam "more luminous than the center of the Sun."

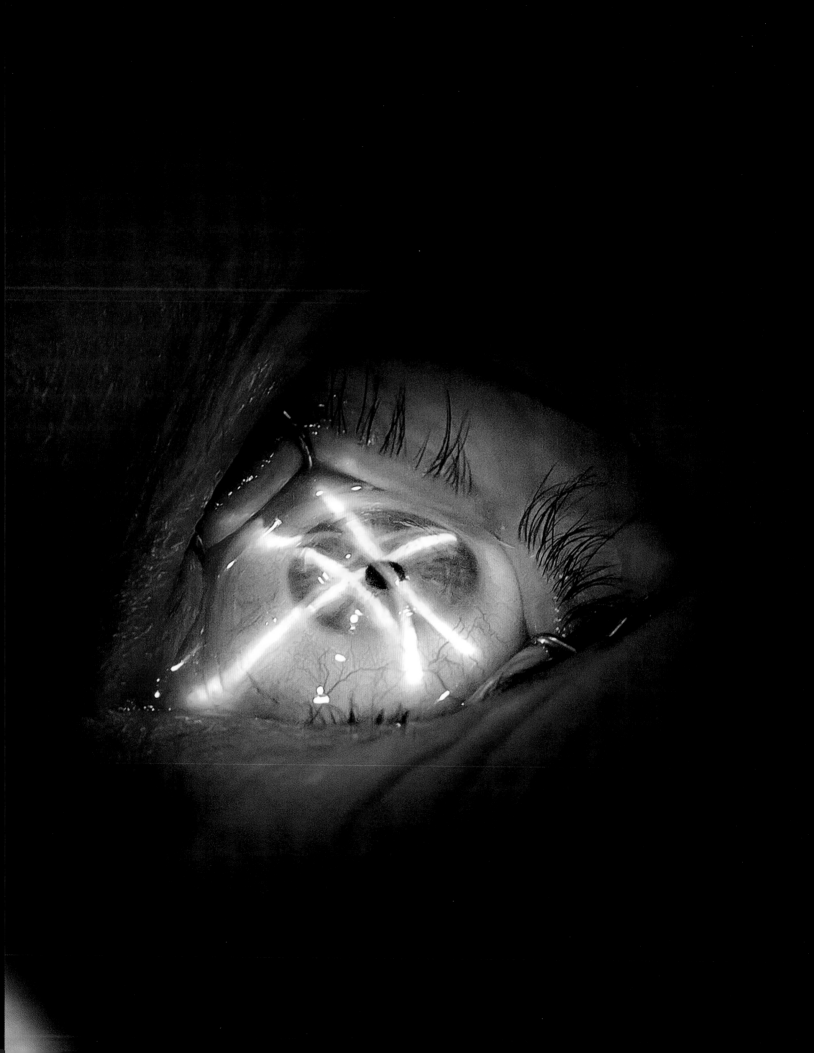

Important industrial applications are cutting systems used for the ultra-precise shaping of the most diverse materials: steel, aluminum, titanium, glass, ceramics, Kevlar, leather, cloth, and plastic. Here the forms are first video drawn to reduce wastage of material as much as possible. Other industrial uses of the laser are welding and marking (laser writing of units on measurement tools, identification codes, trademarks and logos). The last frontier is constituted by quantum computers, in which photon chips exchange information by exploiting photonic impulses rather than electronic ones.

64 Laser microsurgery for the correction of eye disorders. In 1962, two years after its discovery, the laser was already being used in medicine for operations on the retina. It is widely employed in refractive surgery, dermatology, physiotherapy, and urology for the removal of stones with minimally invasive techniques.

65 CO_2 laser for high precision cutting. As it is very powerful, the CO_2 laser is among the most used in the industrial field; it is also used in the medical field, especially in dermatology. Besides solid state and gas lasers, there are several other types: using semi-conductors, organic colorants, metal vapors and excimers.

METROPOLIS

MANUSKRIPT:
THEA VON HARBOU
REGIE·FRITZ LANG

PARUFAMET

Robot

Beyond human capacities

Elektro walked in response to a voice command, moved his head and arms, opened and closed photoelectric eyes, knew 700 words, which he pronounced correctly thanks to a 78 RPM record player, smoked cigarettes, and could even blow up balloons. Elektro was 6.5 feet tall and weighed 265 pounds. Elektro was a humanoid. He lived in Mansfield, Ohio, and his family's name was Westinghouse; the year was 1939. In 1960 he would also act in a Hollywood comedy.

Elektro was one of the numerous robots which made robot history in the twentieth century. Like Eric, Gakutensoku, Elmer and Elsie, and others. But five hundred, a thousand, two thousand years before them, in Alexandria, Byzantium, and Baghdad there were already automata that danced in the theaters, poured drinks and expressed thanks. They were made of ropes, weights and counterweights, pulleys, gear wheels and containers full of grains of wheat to slow gravity. The great Hero of Alexandria, the Arab mathematical engineers and, in the end, Leonardo da Vinci had studied the fundamental mechanisms for a robot to function. They were mechanisms which made it possible to transform energy into movement: joints, cams, differentials, springs, mechanisms imagined, designed and sometimes built.

It was refined mechanics, the same used by watchmakers, who between the sixteenth and the nineteenth centuries created sophisticated timepieces, some of which kept time and some that did not, that astonished sovereigns and common folk alike: the duck stretching out its neck to take food, the waiter serving tea (and clearing the empty cups), the child writing (dipping his pen in the ink), the organ player, moving her bust, her arms, and her fingers on the keyboard, raising her eyebrows and swelling her chest as if she were breathing, to interpret different melodies.

One after another, we have seen Leibniz' binary mathematics, which would be used to develop programs, artificial electrical energy, batteries, transistors to make everything smaller, GPS, microelectronics and the like. In the second half of the twentieth century science fiction has

The poster for the masterpiece of Expressionist and Science Fiction Cinema: *Metropolis*, by Thea von Harbou and Fritz Lang (1927). Among the protagonists: mechanics, cartoons, hallucination, close-ups, the machine man and the android Maria.

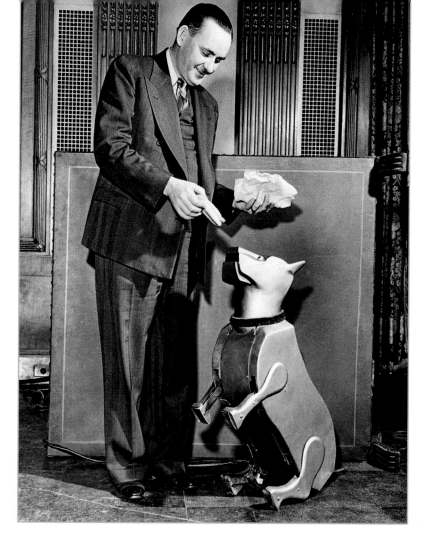

descended to the surface of Planet Earth. And so today we speak of robots, humanoids, bio-inspired robots, cyborgs, androids, replicants and organic androids.

They are all creatures of humankind, the first mechanical and electronic, the more recent completely organic and children of genetic engineering (like *Blade Runner*). The intermediate ones are experiments of outrageous mixtures of organic and artificial, blood, metal and electric impulses.

The keywords are: "sensors", to identify the surrounding environment, map it and relate to the context; "actuators", to act, move, and carry out physical operations; "software", to guide, store data, make mistakes, learn, in short to constantly make choices.

68 The double bassist Lois Kendall plays under the "baton" of Elektro (a prototype robot by Westinghouse Electric) at the World's Fair in Flushing Meadows Park, New York in 1939.

69 Sparko was a 1940 Westinghouse prototype. Here it is seen with its engineer/creator J.M. Barnett. Sparko pretends to smell a hot dog, but really would prefer to be plugged into the nearest socket to fill up with energy.

And so: Curiosity explores Mars. Baxter works on an assembly line together with human workers. OPR picks oranges in Sicily. Grover moves across Greenland to measure the ice sheet. The humanoid HRP-4 dances and sings in Japan. Crabster CR200 has six legs and explores the seabed in Korea. Robonaut 2 lends a hand to the astronauts on the International Space Station. The Korean Mahru-Z helps with housework. PakBot inspects the Fukushima nuclear power station. Geminoid-DK is the perfect replicant double of a well-known Danish IT specialist. Pepper recognizes human facial expressions and, therefore, emotions. Disarmadillo identifies land mines in war zones. Coman moves with the same agility as a gymnast. ICub is a child robot, ten years old and 3.3 feet tall: his hands are identical to human hands, his entire body is sensitive to warmth and touch, but above all he learns; his brain consists of the computers which fill a whole room. There are 25 iCubs around the world, from Illinois to Osaka—each one growing with its family of scientists.

There are other keywords which are already disturbing, like self-learning, experience, feelings, creativity. But perhaps what is most fascinating and frightening, what really straddles the red line, which includes almost everything, what is concerned with ethics, philosophy, the very essence of the human being, is "artificial intelligence." As Stephen Hawking always said: "I fear that the development of full artificial intelligence could spell the end of the human race. If people design computer viruses, someone will design an artificial intelligence that improves and replicates itself. This will be a new form of life that outperforms humans."

The curious gaze of Pepper, a robot of the latest generation (2016) from SoftBank Robotics, Tokyo. It does not perform work, but moves autonomously, takes part in conversations, and keeps people company. All for 1800 dollars.

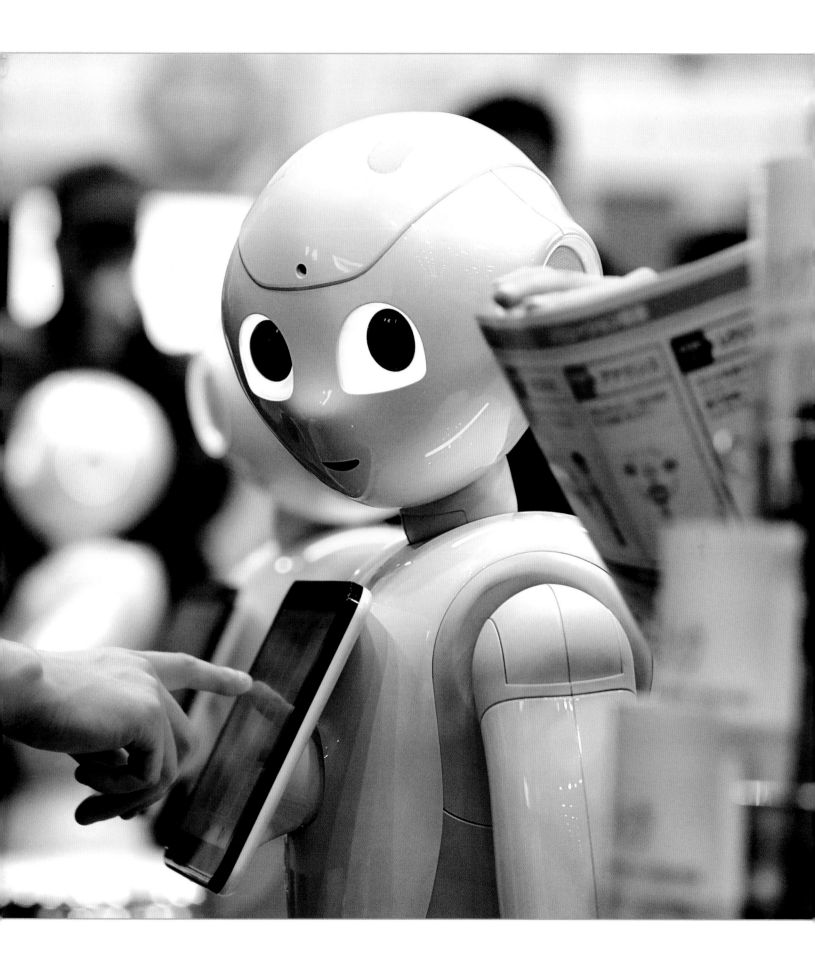

DAILY LIFE

Typewriter
Light Bulb Zipper
Refrigerator
Ballpoint Pen
Credit Card

They are all objects that today we take for granted, even those no longer used, like the typewriter. In some way, their advent has brought about fundamental changes in family, social relationships, and work. Chronologically, the first revolution was the light bulb, providing electric light in the darkness of the cities, streets, homes, cafés, churches, workshops, and warehouses. Before then, there had been candles, of course, but theatrical scenery completely illuminated by color spotlights amazed everyone for many, many years. Shortly afterward, at the end of the nineteenth century, food preservation companies adopted the refrigerator. However, it was between the '20s and the '50s that it came into homes, and here it set off epochal change, especially in women's lives: electrical appliances created time for them to dedicate to themselves and to work outside the home. Still today, advertising is mainly directed at them, with images of appliances that are increasingly efficient, extraordinary, and even beautiful. The zipper was a small, albeit fundamental, element that marked the change from rigidly conservative fashion, above all for women, to freedom from laces, corsets, and complicated ways of dressing.

Two writing instruments burst on the scene to replace nibs, inks, and fountain pens. The ballpoint pen, which for a long time was opposed by primary school teachers and lovers of calligraphy, in the end won its place in the pockets and penholders of professors, managers, clerks, cashiers, and salespeople. Although typewriters were destined for a short life, for decades they were everywhere in newspaper editorial departments, on writers' desks, and also in offices the world over, which were full of typists doing hundreds of words per minute.

The last in our review is the credit card, a universal instrument as long as one has money to spend and the desire to spend it (without even seeing it). A step forward toward abstraction, toward a world composed of invisible electrical impulses. In short, of all the inventions that revolutionized the last century, perhaps it is the simplest ones that most radically transformed our daily life. After all, it is from daily life that revolutions have always begun.

In an image from the beginning of the twentieth century, one of the first models of typewriter by Remington and Sons; it was similar to the sewing machines in which the US company had specialized. The pedal controlled the carriage return. Production began in 1870.

Typewriter

Many patents and records in a brief century of glory

Like so many great inventions that have revolutionized society, the typewriter had more than one inventor. Mechanical devices to replace handwriting and make it faster were perfected over the centuries, and before achieving a standard model scores of patents were necessary. Among those who could lay claim to the original idea, there are a typographer from Venice (Francesco Rampazetto, who in 1575 built a machine for the blind), a hydraulic engineer from London (Henry Mill, who patented his typewriter in 1714), a gentleman from the Tuscan Garfagnana region (Pellegrino Turri, the inventor of carbon paper, who in 1802 made his *preziosa stamperia* [invaluable mechanical typing machine] for a blind female friend), a lawyer from Novara (Giuseppe Ravizza, who called his invention *cembalo scrivano* ["harpsichord scribe"], patented in 1855), and even a Brazilian priest and a South Tyrolese carpenter.

We owe to the Danish Christian minister, Rasmus Malling-Hansen, the first machine to be offered for sale, in 1870: the keyboard was a sort of ball and in no way resembled a typewriter. However, the first model with the classic QWERTY keyboard, which we find on today's computers, was perfected by a group of American inventors and went into production by E. Remington and Sons, which already produced weapons and sewing machines. Later, the world's largest manufacturer became IBM, which developed the electric typewriter, while the record for design goes to the Italian Olivetti: the *Lettera 22* by Marcello Nizzoli (1950), the *Valentine* by Ettore Sottsass (1968), and the electronic ETP 55 by Mario Bellini (1987) are only the most famous of about ten Olivetti machines exhibited in the permanent collection of the MoMA.

After hundreds of millions had been sold, the typewriter went into rapid decline after existing for just a century. It has now been completely replaced by word processing programs and relegated to modern antiques markets, but in human history it remains a fundamental object for writing in all its contexts, and in the evolution of social and economic relations in the twentieth century.

76 An American typewriter model from the end of the nineteenth century (top), with an external roll and vertical mechanical move-ment. The internal mechanics of the Olivetti M1 (bottom), produced in 1911. The company had been founded three years before, in Ivrea, by Camillo Olivetti.

77 A typist tries the *cembalo scrivano* ("harpsichord scribe"), the wooden ancestor of modern typewriters patented in 1855 by the Novara lawyer Giuseppe Ravizza. A model is preserved today in the Museo della Scienza e della Tecnica di Milano (Milan National Science and Technology Museum).

78 The Friden Flexowriter was one of the first electric typewriters, designed by E. Remington and Sons in the '20s and then developed by IBM, which produced it until the '70s. When linked to calculators, it became a predecessor of the modern computer.

79 Two famous Olivetti models, the *Lettera 22* designed by Marcello Nizzoli, winner of the Premio Compasso d'Oro in 1954, and the *Valentine* by Ettore Sottsass and Perry A. King. Both models belong to the permanent collection of the MoMA in New York.

Light Bulb

Thomas Edison, the father of home and city lighting

The panoramic views from space reveal to us our planet dotted with whitish, yellow and golden lights, a network that becomes denser near metropolises, sparser in the countryside and the deserts. Seen from above, the Earth is like a star-studded sky created by humankind.

If we imagined the same view at the end of the nineteenth century, the Earth would appear dark, lit only by a few oil lamps, flamelets that burned oil, and a few arc lamps creating light with an electrical discharge. In contrast, the stars in the sky would shine more brightly than today.

Thus it appeared to Joseph Wilson Swan from his house in Gateshead, in the north of England, when his experiments were only attempts by a chemist grappling with electricity. In 1878, when at last Swan switched on the first incandescent light bulb with a carbon filament, it was the first time that a house had been lit by electric light. The day no longer ended with the setting sun; the night was no longer dark, and it extended until dawn the next day. Swan registered the patent and began to produce light bulbs. In 1881, the stage of the Savoy Theatre, in London, would be illuminated as day, in what would be the first public building to enjoy this privilege.

Across the Atlantic, on the other side of the world, someone else was following the same path. In 1879, Thomas Alva Edison registered his umpteenth patent (one of 1,093 in his name over his long career): the incandescent light bulb with a high-resistance filament. In contrast to the carbon bulb, it did not emit gas inside the bulb, and thus did not create soot; it lasted longer, consuming less energy. As a result, Swan introduced improvements to his product, and Edison sued him for patent infringement.

Thomas Alva Edison in the laboratory in Menlo Park, New Jersey. He was not the inventor of the light bulb, but he perfected its technology and transformed it into an everyday object, in the home, in factories, in the streets. It was an easily produced device, simple to install, and low in energy consumption.

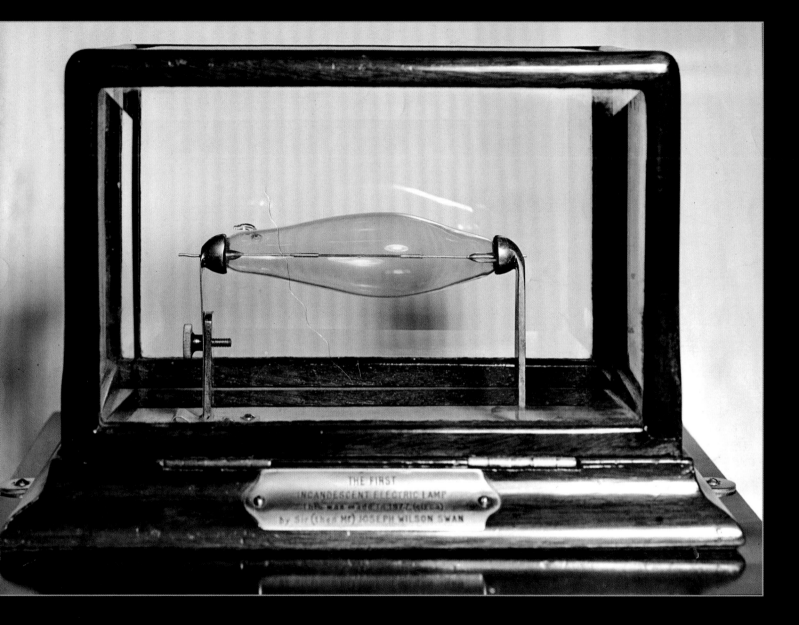

The first incandescent light bulb, created by Joseph Wilson Swan in 1878, comprised a vacuum-sealed glass bulb containing a carbon filament. In 1888, carbon was replaced by tantalum, a stronger metal. Swan was also a pioneer in production and opened the first light bulb factory in Newcastle four years after his invention.

Joseph Wilson Swan at age 77. He was a philosopher, inventor, and a student of chemistry and physics. Besides the light bulb, he invented carbon paper, a bromide photographic paper, and a cheaper photomechanical procedure than those existing at the time.

Light Bulb

EDISWAN
POINTOLITE

As always happens, "when you can't beat them, join them": Edison-Swan was founded, which increased trade between Europe and the United States.

As early as 1910, there was the first tungsten-filament light bulb—the one that illuminated our houses for over a century before giving way to LED and low consumption bulbs.

From darkness to constant, lasting light. A revolutionary invention. It is no accident, in fact, that in the collective imagination the light bulb represents the idea, intuition *par excellence*.

84 A standard light bulb by Ediswan, the British producer of incandescent light bulbs and other electrical products, formed by the merger between Swan United Electric Company and Edison Electric Light Company in 1883.

85 Advertisement for Edison light bulbs from 1924. Almost by magic, a smiling female figure holding a light bulb appears from the darkness as she illuminates the city of Milan.

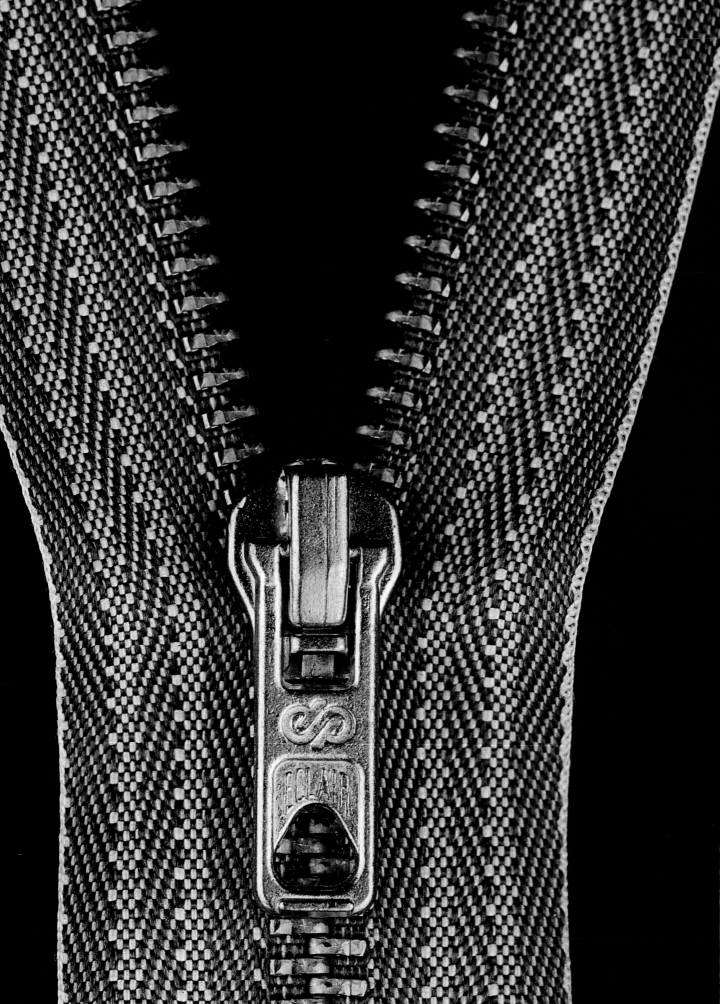

Zipper

The zip is quick!

A slight, discreet, little gleam that betrays a new presence, hidden on the pockets of the money belts of American soldiers leaving for the First World War. The invention was the work of the Swede, Gideon Sundbäck, employed by Whitcomb Judson's Universal Fastener Company.

In fact, by perfecting the system of hooks and eyes created by Elias Howe (the inventor of the sewing machine), in 1893 Whitcomb Judson had deposited the patent for a "separable safety closure", or clasp locker, to replace shoe and boot laces. However, because of the unreliability of the mechanism and its excessive production time, he had not obtained the success he had hoped for. Judson had not given up, and he subsequently founded the Universal Fastener Company, working ceaselessly to improve his creature.

The turning point came in 1906, when a young Swede presented himself to the company, laden with designs, hopes and *bratwurst* from Germany, where he had studied to be an engineer. Gideon Sundbäck quickly managed to make a good impression on the head designer Peter Aronsson, who was also of Swedish origin, and to win the heart of his daughter, Elvira. He was thus tasked personally with Judson's invention. In 1913, he invented the first "modern" zipper, in which small metal teeth took the place of the hooks and eyes and were placed in cloth strips, so that they could be sewn onto fabrics. He quickly moved from soldiers' belts to rubber galoshes: they would be called Zipper Boots and the name by which we know this device today derives from Sundbäck's creation.

Once the focus of a bitter contest between the traditionalist supporters of the button and the progressives of the zip, today the zipper is seen on backpacks, clothes, and bags. It has now taken the place of its bitter rival even on those Levi's jeans that made their appearance in 1873 in the fashion world, ready to revolutionize it.

Zipper

In the '30s the fashion designer Elsa Schiaparelli introduced it to *Haute Couture,* freeing generations of women from the need for laces and buttons on evening and other dresses. The zipper was transformed into a visible, colorful accessory, and proclaimed its independence.

Precisely in those years, French stylists went into battle in the "Battle of the Fly", claiming that zippers were superior to buttons even on men's jeans. The microscopic metal tongue thus achieved its place in the world, with all its advantages. It was even praised by the magazine *Esquire*, which in 1937 recognized that it avoided "unintentional and embarrassing disarray."

A female worker on a machine for zip production in 1959. The first factory was opened in Switzerland by Martin Othmar Winterhalter, who in 1923 acquired the zip patent from Gideon Sundbäck. He thus launched the mechanization of a manufacturing process that enabled the zip to conquer the world, even in Asian countries where three letters (YKK) would become synonymous with "zip".

WHITE MOUNTAIN

WHITE MOUNTAIN GRAND

REFRIGERATORS

· 1903 ·

MAINE MANUFACTURING COMPANY
NASHUA · N·H · U·S·A
ESTABLISHED – 1874 ·

Ice-houses in general were normal cupboards with compartments: a block of ice was inserted into one, and it cooled the other shelves where the food was kept. This 1903 poster advertises a model by the Maine Manufacturing Company, New Hampshire.

Refrigerator

Ice everywhere, for everyone!

The use of fire represented a turning point in human evolution, but also the "conquest of cold" was an important achievement in daily life and other contexts.

For centuries, refrigeration had been one of the fundamental systems for preserving food. Before modern industrial and chemical techniques, it was already known that low temperatures slow the deterioration of foodstuffs, avoiding risks to our health and reducing waste. For many years, ice was used, and caves, buildings and containers were filled with it to store food: these were generally called "ice-houses". Then in the nineteenth century, thermodynamics came into the equation and everything changed.

This new branch of physics conceived a cycle in which a gas is allowed to expand and be compressed, and is able to take heat from one space and transfer it to another. This is the basis of the first artificial refrigerating machine, the first refrigerator, which is attributed in 1834 to Jacob Perkins, but it had several previous prototypes and very many later developments. The gas—ether and ammonia in the first machines—was allowed to expand through tubes: in order for its volume to increase, it needed energy, which it absorbed from inside the refrigerator, lowering its temperature. Then the same gas was recompressed, releasing energy into the surrounding space.

The actual use of the new invention, whose revolutionary practical usefulness was soon obvious, began at the end of the century, with the introduction of industrial machines. Only in the '20s did refrigerators come into the home, while they boomed in the '50s, when they began to be produced on a large scale by large corporations like General Motors, Bosch, and even FIAT. They even became common on means of transportation like trucks, vans, ships, trains, and fishing boats. The refrigerator preserves food, but also lengthens the life of other perishable products, like pharmaceuticals and industrial chemical products. Once it had entered daily life, its evolution continued: it continues relentlessly also in the third millennium.

92   The Frigidaire prototype by Delco Electronics, a General Motors company, was produced in 1921 in Dayton, Ohio. Its gas-compression cooling system made ice obsolete. Among the first refrigerants were ammonia (toxic) and Freon (highly polluting); today neither one is used.

93   The B-8 model (with cooler) from General Electric, an American company that led the way in the production and sale of new refrigerators. In 1927 the Monitor-Top, an ancestor of the model in the photo, was the first to become widespread in middle-class North American homes, well before the boom of the '50s.

Refrigerator

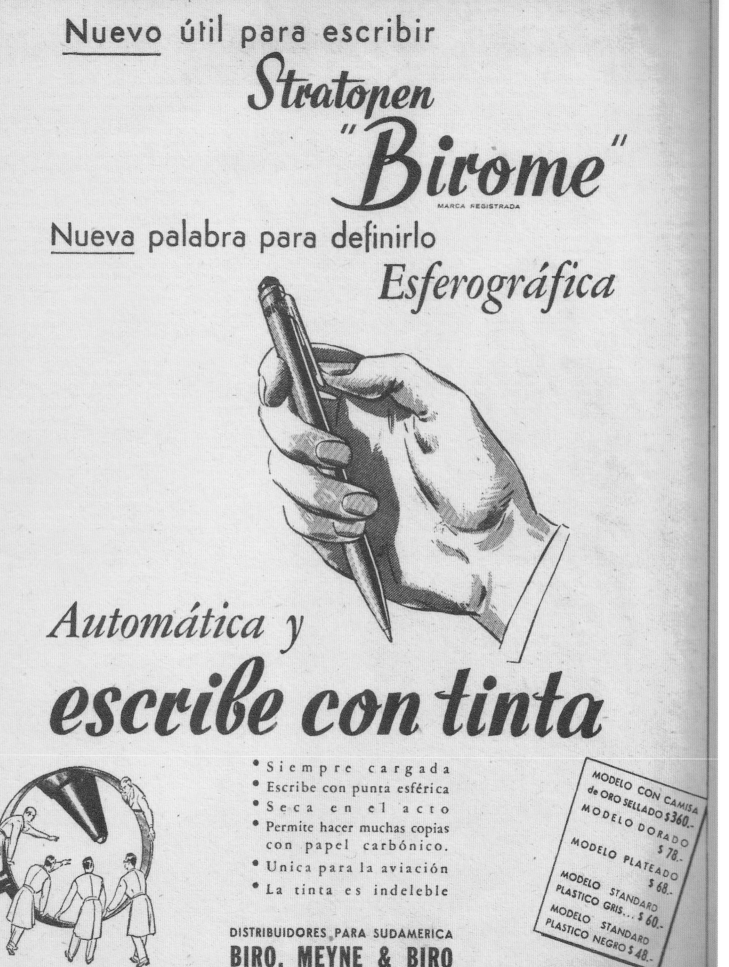

Nuevo útil para escribir
Stratopen
"*Birome*"
MARCA REGISTRADA

Nueva palabra para definirlo
Esferográfica

Automática y
escribe con tinta

- Siempre cargada
- Escribe con punta esférica
- Seca en el acto
- Permite hacer muchas copias con papel carbónico.
- Unica para la aviación
- La tinta es indeleble

MODELO CON CAMISA de ORO SELLADO $360.-
MODELO DORADO $ 78.-
MODELO PLATEADO $ 68.-
MODELO STANDARD PLASTICO GRIS... $ 60.-
MODELO STANDARD PLASTICO NEGRO $ 48.-

DISTRIBUIDORES PARA SUDAMERICA
BIRO, MEYNE & BIRO
Alsina 633 - Buenos Aires - U. T. 34-9958

Venta en todas las casas del ramo

In 1945 the magazine *Leoplán* advertised Birome, a ball-pen marketed in Argentina, an acronym combined the name of the inventor, Biró, with that of his friend Meyne.

Ballpoint Pen

A Hungarian artist and an Italian-French baron:
the origins of an empire

June 15th, 1938, *Wall Street Journal*: "A simple but remarkable invention came into a world about to be convulsed by death and destruction." It speaks of the first prototype of the ballpoint pen. The ink must not blot the page—this was the mantra. László Bíró, a Hungarian from a Jewish family, was at work with his brother György, a chemist. The two of them had already had a brilliant idea which would improve writing: the use of denser inks, starting from those used in newspaper rotary presses. The problem remained of how to make the ink run properly from the pen to the sheet of paper.

The idea had surfaced some time before, in Budapest, when Bíró had observed a child throw a glass marble into a puddle: it left a sharp and regular liquid trace. Driven by an atavistic obsession with always having clean hands, Bíró thought of a writing instrument that could leave an equally precise trace without leaving blots: the ball pen.

Bíró was a student of medicine, painter, journalist and an amateur inventor. He made his fortune with hypnosis and stood to make even more with this brilliant discovery.

There were endless obstacles on the road to success: the Second World War, the anti-Jewish persecutions, and the entrepreneur Marcel Bich. Forced to emigrate, first to Paris and then to Buenos Aires, Bíró patented a new prototype of his pen and registered it in Argentina.

It was 1943 and production was fostered by the need to write at high altitudes. The British Royal Air Force made an order for three thousand pens and the biro began to spread around the world.

Bíró decided to sell the patent for 2 million dollars to Marcel Bich. The Italo-French baron perfected the design and marketed the ballpoint pen, making it cheaper. The Bic was born: it was a simple and effective name. From 10 dollars in 1946 (equivalent to 100 dollars today) the price fell to 19 cents in 1959, and the Bic became the most famous and most used pen in the world. Its inventor Bíró dedicated himself again to painting and to his inventions; he died in Buenos Aires in 1985.

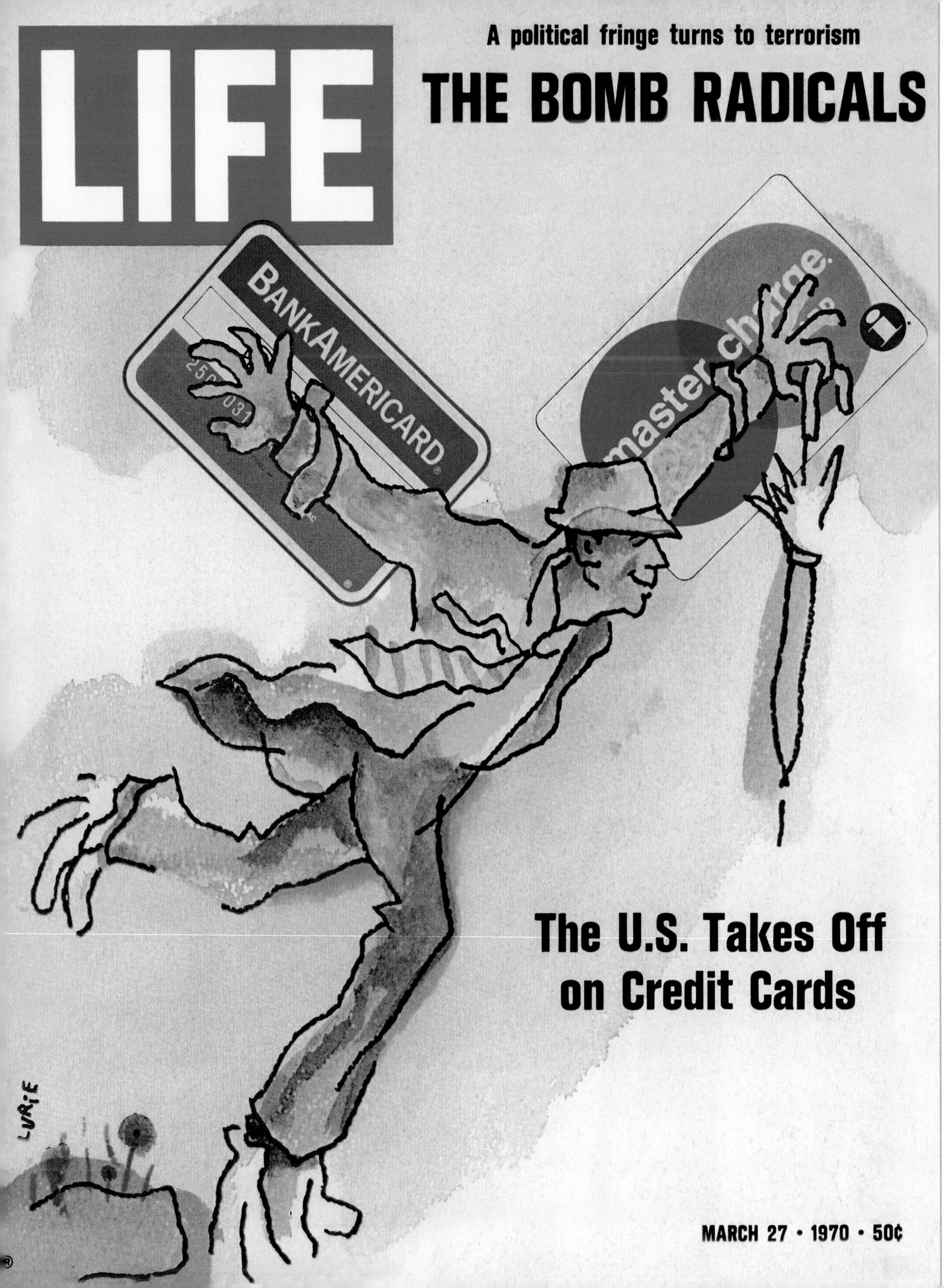

LIFE

A political fringe turns to terrorism

THE BOMB RADICALS

BANKAMERICARD.

master charge.

The U.S. Takes Off on Credit Cards

LURIE

MARCH 27 · 1970 · 50¢

The figure on this cover of *Life* from 1970 (illustrated by Ranan Lurie) happily takes flight with the support of the two pillars in the burgeoning credit card business: BankAmericard and Mastercharge, respectively the first and last cards in order of evolution.

Credit Card

The magnetic card that made money virtual

It was a September morning like any other. To the sound of the alarm, routine came in through the windows of the sleepy town of Fresno, together with the first, timid rays of sun. Pancakes, fresh-squeezed juice, the newspaper and the mailbox . . . and what a surprise for its 60,000 inhabitants to find an envelope from Bank of America! Inside it, there was a small plastic card with a letter of presentation, describing the use of the BankAmericard and inviting citizens to try it—it was the latest in money, it was the "credit card".

By convention, this event in 1958, which then went down in history as the "Fresno Drop", marked the beginning of a new era for the credit card, that of its widespread possession.

Actually, the ancestors of the card had already peeped out at different moments in history; in the eighteenth century, a furniture seller had introduced the installment plan for his customers. At the turn of the century, sellers had begun to distribute, to their best customers, small medal-like objects with holes, with the name and account number printed on them, to be presented with payment. However, every store had its own "medal" and very soon customers' pockets were full (and their wallets empty).

It was then that Frank McNamara, a well-known New York entrepreneur, had the idea in 1949 of creating a card that managers and businessmen could use to pay on credit in a series of restaurants and hotels in New York which were part of the scheme: the Diners Card.

Its restricted circulation, however, did not facilitate its use, so Bank of America decided to adopt the "Fresno stratagem". Following the Bank of America's granting of licenses, more and more banks adopted the credit card, and in 1966 the fourteen banks in California united in a consortium to formulate regulations. Thus Interlink, VISA's predecessor, was founded, which was soon opposed by MasterCharge, later MasterCard. Money became virtual.

In the '70s, the magnetic strip was introduced and shop owners' confidence in the new invention increased, helping it to spread like wildfire.

Today, our wallets are full of cards of every type and color, from ATM cards to debit cards. They are instruments that undoubtedly promote consumption and lessen awareness of spending.

TRANSPORTATION

Airship

Automobile

Airplane

A Boeing 747 flies at 29,500 feet altitude at 600 miles per hour, covering a distance of 8,000 miles without refueling, and carrying more than 500 passengers (and their baggage). A sedan with four passengers and baggage for a month's vacation travels along the highway at 80 mph, and we do not even think about it.

Today it seems incredible, and yet for thousands of years the horse was the unit of measurement, together with mules, camels, carts, carriages and not much else. No more than 30 miles a day, 6-7 hours' travelling, a half hour to rest. Carts were much slower, as were mules, which were mostly used in the mountains; these carried several pounds of goods in the packsaddle. Only messengers conveying royal or military messages traveled faster: they mounted a new horse at staging posts every 20 miles, and by riding from dawn to sunset they could cover up to 180-220 miles a day.

For centuries, nothing changed, except for the number of bridges and roads, which were well built and well maintained in the great empires and in wealthier and more astute kingdoms, but left to themselves in other cases. All this was apart from crossing mountain chains, bad weather, landslides, floods, problems with bandits and customs men, and so on.

The seas and oceans were no better: a possible average was from 30 to 50 miles a day. But even more than on land, the winds and weather were arbiters and protagonists of the situation. Days of waiting for favorable wind to sail, storms, delays in harbors. And the cargoes were limited by the necessity to take water, provisions, and other necessities for the crew, on voyages that lasted weeks.

And then, suddenly, there was the revolution of the nineteenth century, with the steam engine on railroad locomotives and the great ships crossing the Atlantic; at the turn of the century, it was improved by the turbine. Despite the Titanic disaster, from that moment on there was a race toward power, speed, enormous size (the airship Hindenburg was as long as two soccer fields and as high as a 15-story building). The internal combustion engine, the diesel engine, propeller engines, jet propulsion. Automobiles, motorcycles, trucks, planes, steamships, container ships, oil tankers. They are still working on space transport, but it will not be long.

Airship

Will the giants of the air fly again?

Dusk, May 6th, 1937, New Jersey. The crew of the airship LZ 129 Hindenburg threw the mooring cable to the ground. Shortly before then, the airship had sailed through a violent thunderstorm that had charged its metal skeleton with static electricity. When the cable touched the ground, the circuit was completed and produced electrical discharges and sparks. The envelope of the Hindenburg contained 53 million gallons of highly inflammable hydrogen, and in a few seconds it was transformed into a gigantic torch 800 feet long and 155 feet high. For the first time, a tragedy was broadcast live on the radio and shown in newspapers by the photographers present, who were waiting for the largest ever creature of the skies. That fire meant the end of airships and their war against the airplane.

In France in 1850, they had built the first airships by applying to hot air balloons—which could only ascend and descend—a few modifications, especially a "motor" able to "steer" them independently of the wind. But the history of airships is inextricably linked to a works in Southern Germany, the *Luftschiffbau Zeppelin GmBH*, simply Zeppelin for the rest of the world. After the First World War, for almost 20 years, it built airships for the United States and for Germany.

With 50 passengers and a crew of 60, the airship Hindenburg (LZ 129) here flies over Manhattan. In July, 1936 it set the record for an Atlantic crossing: 5 days, 19 hours and 51 minutes.

They had a new design, of the so-called "rigid" type; that is with an aluminum frame supporting many bulkheads containing the gas. They had steering in the stern, several powerful engines, and the "gondola" (the external cabin for the crew), while the passengers were housed in a great internal cabin in the body of the airship: a luxury cabin with a restaurant, bar, lounges, and many rooms. Between 1928 and 1937, the LZ 127 Graf Zeppelin, the most famous of the fleet, flew almost a million miles, crossed the Atlantic hundreds of times, and was even the protagonist in the circumnavigation of the globe.

102-103 In 1935, in the enormous hangar of the Zahnrad-fabrik Friedrichshafen the outer framing is mounted on the rigid structure of the airship LZ 129, the future Hindenburg.

103 Top March 1936: the elegant promenade deck for guests on the Hindenburg.

In fact, it was flying over the Atlantic from South America to Europe when the captain was informed of the Hindenburg disaster. It landed in Germany on May 8th, 1937 and from then until the '90s no more Zeppelins took off.

At the end of the twentieth century, airships reappeared, to be used for advertising or for tourist flights. Today, they are being considered for use as cargo transportation or as space bases on the edge of the Earth's atmosphere. Perhaps a return of the airship is yet to come.

104   The airship Italia—commanded by the Italian general Umberto Nobile—moored in the Svalbard in 1936, ready to sail toward the North Pole. The mission, conceived by Nobile, had the objective of leaving a tent with a group of scientists at the Pole, but it ended in tragedy. Part of the crew was rescued after 49 long days. Another part of the crew and the structure of the airship disappeared forever.

105   The tragedy of the Hindenburg at Lakehurst lasted a few seconds. It was 6:30 a.m. on May 6th, 1936. There were 36 fatalities. The photo is one of the most famous and dramatic in the history of aviation.

The German engineer and automobile manufacturer Karl Benz driving the tricycle he had invented. The Patent-Motorwagen (1886) was the first auto in history with an internal combustion engine.

Automobile

When horsepower replaced horses on the roads

"Certain fads can prove dangerous": if you combine a woman ahead of her times, a passion for showy foulards, and a convertible, it's a 100% certainty. This is how Gertrude Stein commented with undisputed cynicism on the death of Isadora Duncan, the famous (and sometimes troublesome) American dancer, who was strangled by a scarf caught in the wheel of her luxurious racing Bugatti.

But in 1927 the car was still a "fad" for the few, and not everyone appreciated its revolutionary importance. But Jules Bonnot did. He was a French anarchist who, as early as 1911, carried out the first of a series of audacious robberies in Paris with his band in a stolen Delaunay-Belleville. They were certainly robbers, but they had a very clear intention: to strike at the heart of capitalist society—the banks—with the product of capitalism—the automobile.

In the first decade of the century, the car had already been perfected, but it was still a luxury product, for the few. Apart from some steam engines powering carts, the nineteenth century was the century of the railroad, and the auto was still little more than an idea: the Obéissante and the Mancelle by Amédée Bollée were exhibited at the Universal Exposition of 1887 as a steam car. Only at the end of the century did the auto begin to impose its personality, when competition with the middle-class, horse-drawn carriage became reality, with the improvement in the four-stroke internal combustion engine by the German Nikolaus August Otto and the subsequent popularity of the gasoline engine.

The first plants were created: in Germany, Benz & Cie of Mannheim was prominent. Launched by the engineer Karl Benz with the help of his wife, Bertha, it attracted the attention of the press with the first public launch of the Patent-Motorwagen. It's curious that despite the many subsequent jokes about women drivers, the first prototype was driven on public highways by a woman, who covered the considerable distance of 65 miles km to see her mother. In France, at the turn of the century De Dion-Bouton became the first manufacturer in the world. In Italy, the first car manufacturer was Enrico Bernardi's Miari & Giusti of Padua (1894); FIAT, in Turin, launched industrial production in 1899. In the meantime, Daimler's Mercedes 35PS was marked by its new refined and captivating lines, and for its top speed, then unthinkable, of 43.5 mph. The speed of 60 mph was achieved for the first time in April, 1899 near Paris by a car whose name was its destiny: Jamais Contente. The first auto races thrilled not only the competition, but also ordinary people. One day, after the Second World War, the symbol of car racing would be founded: Ferrari.

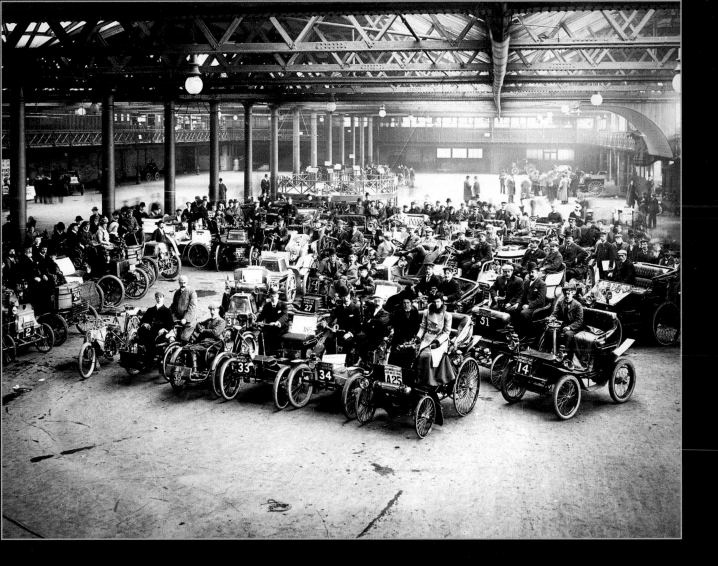

108 London, 1900. Group photo participants in the 1,000 Mile Trial, from the album in the collection of Charles Stewart Rolls, the British driver and aviator who, together with Frederick Henry Royce, founded Rolls-Royce Ltd (1906). The winner of the competition would, in fact, be Rolls in a 12 HP Panhard.

109 The Targa Florio was one of the oldest car races in the world. It was held in Italy from 1906 to 1977, almost always following the curves of the Madonie mountains in Sicily. In this photo we see Vincenzo Trucco in the 1908 race driving his Isotta Fraschini, in which he would win the race.

Already some were thinking about mass production to lower costs and make the car accessible to all. The first entrepreneurs to believe in it were Americans. While Ransom Eli Olds, the pioneer of the assembly line, was not successful, in little more than a decade Henry Ford managed to reduce costs dramatically and conquer the American market with his black Model T.

The interwar years created the myth of the automobile: a status symbol for men and a symbol of emancipation for women, it evoked the luxurious life of the Hollywood stars and the European *bel monde*. The names were legendary: Rolls Royce, Isotta Fraschini, or the Packards with which Al Capone's gangsters terrorized Chicago. The "people's car" featured strongly in the populist speeches by dictators like Hitler and Mussolini and, in the '30s, it would be Volkswagen in Germany and the Balilla in Italy. The boost in production after the Second World War would cause an explosion in all the West.

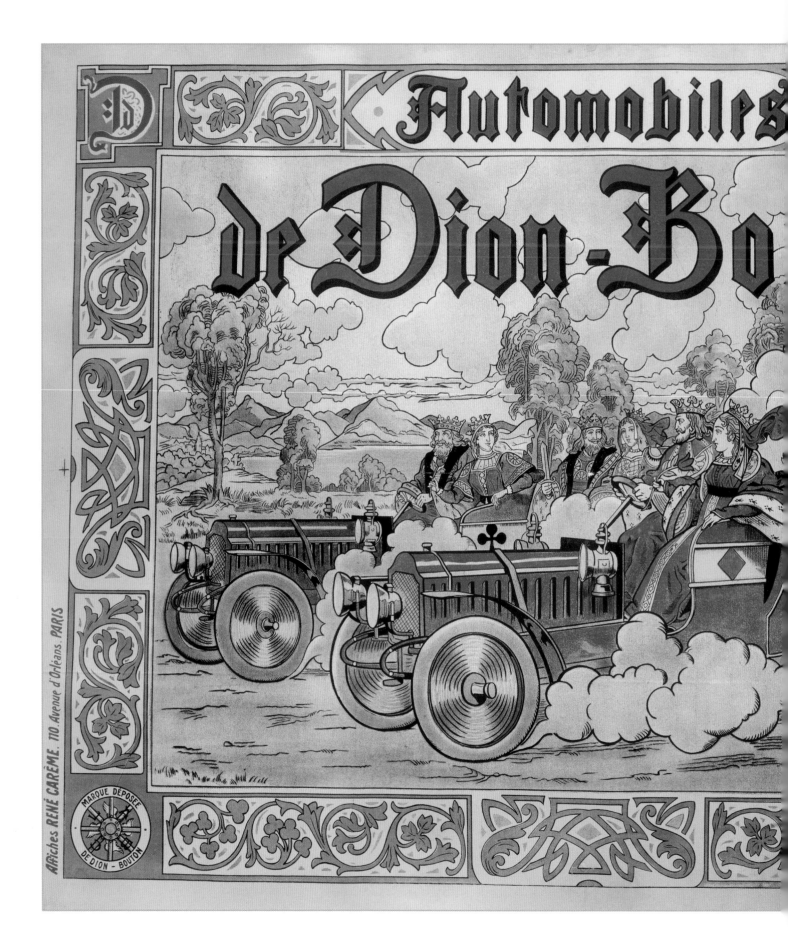

In the '60s and '70s it was the auto industry that drove the economic boom: the car was now king and almost every family owned one. Some cars made history: Elvis Presley's pink Cadillac, the symbol of the American dream, and the Chevrolet Corvette and other enormous cars circulating on American highways; in Europe, timeless cars like the Volkswagen Beetle/Bug, the Citroen 2 CV, the FIAT 500, the Morris Mini. In Japan the star of Toyota would soon rise.

For the dominance of the car to be questioned, it would take the oil crisis, an increase in air pollution, and congestion in large cities. But now the car has left its mark on the habits and behavior of everyone, and it will be difficult to do without it.

110-111 Advertising poster from the '20s for De Dion-Bouton: the king and queen of playing cards compete in cars decorated with the four suits: hearts, aces, spades, and clubs.

111 Top Henry Ford personally presents the New Ford Model T in the United States, ca. 1900.

112   In the advertising poster for the FIAT 508 Balilla there is a boy in Fascist uniform ready to hurl a stone. The drawing pays homage to Giovan Battista Perasso, nicknamed Balilla, who, with a stone, set off the Genoa revolt against the Austrians in 1746. This FIAT model was presented at the 1932 Milan Motor Show, and it became the most popular for the Turin company. It covered 60 miles on only 2 US gallons of fuel.

113   On May 27th, 1938 the Chancellor of the Reich, Adolf Hitler, inspects a new "people's car" at the Fallersleben plant, in Germany. The car was designed for a production of 6 million. On the left of Hitler we see the car designer Ferdinand Porsche, the founder of Volkswagen.

Airplane

Otto Lilienthal's glider flying down a slope near Berlin on May 29th, 1895. From that day, the low hill would be named in his honor as Fliegeberg (flying mountain) and later Lilienthalberg (Lilienthal's mountain).

The Wright brothers' dream takes flight

They are 246 feet long, have a 230 feet wingspan, weigh 500 tons at full capacity, can hold 500-600 passengers and reach over 600 miles per hour at top speed, and have a fuel distance of almost 10,000 miles (about halfway around the world): the latest commercial aircraft are monsters in metal and plastic that fly through the air effortlessly.

The airplane was an invention that left its mark on the twentieth century with a relentless series of technological and engineering achievements, from the first propeller planes, to jets with reaction-type turbines, to the beginning of modern commercial aviation. Airships were already on intercontinental routes, but the challenge was to fly a heavier-than-air craft.

That challenge was met thanks to the brilliant intuitions and extraordinary courage of two brothers with a passion for mechanics and bicycles. Wilbur and Orville Wright built bicycles in Dayton, Ohio. They left their studies and opened a store, the Wright Cycle Co. No one could have imagined that a few years later they would change the world. In fact, bicycles were not the only "machines" to attract their attention.

These were years in which many inventors were trying to build machines, like gliders, that were capable of flying. The most famous was the German Otto Lilienthal, who had already made more than 2,000 attempts to fly, but who died in 1896 following a flying accident. The news of the accident also reached the Wright Brothers in Ohio.

Wilbur and Orville were convinced that the principles leading Lilienthal to make progress with his glider were a good starting point. The problem was to succeed in controlling the movements of the glider perfectly before building an engine able to keep it in the air for longer. Fundamentally, for the Wrights, the principle maintaining the motion of a flying machine was not dissimilar to that moving the bicycle. They understood that the control of an "airplane" involved three different axes: movement around the transverse axis, called pitch, movement around the vertical axis, or yaw, and movement around the longitudinal axis, roll or lateral movement. The last was the most difficult to control: birds' movement of wingtips in flight was not easily replicable on a glider.

The Wrights continued their engineering experiments for two more years in the attempt to solve the problem. In 1901, they built a glider large enough to carry a man in a prone position. In the summer of the

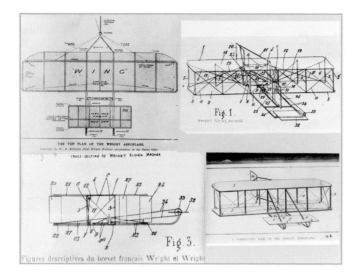

same year, they looked for a place to test it. They need-ed a continuous wind to support the glider in flight: the beach of Kitty Hawk, in North Carolina, offered the best conditions, and so Wilbur and Orville left to test the prototype.

It was Wilbur, the elder of the two, who piloted the glider on that occasion. When the glider gained height, it reached an altitude of 400 feet, but it soon went out of control and crashed: Wilbur was seriously injured. Some time afterward, new flight tests led to an important discovery: to maintain control of the glider, the rudder can compensate for the movements of the wings. The Wrights designed a new rudder—no longer fixed, but mobile—which turned out to be the solution.

Up to then, the wing design used by the Wrights was the same as in Otto Lilienthal's model. The broth-

116 Top   The design of the Wright brothers' airplane.

116-117   The first flight of the Flyer I in Kitty Hawk, North Caro-lina, on December 17th, 1903.

ers decided to construct a small wind tunnel to redesign the wings and check their calculations. The tests revealed that a wing with an airfoil-shaped cross section, thicker in front and thinner behind, reduced the air pressure on the upper surface and increased it on the lower surface, thus maximizing the glider's lift.

In the summer of 1902, the Wrights returned to Kitty Hawk, ready to test the new prototype. This time, everything worked perfectly. In 1903, they perfected Flyer I, a motorized version of their glider. On December 17th, 1903, the airplane was officially born: a small, curious crowd gathered on the Kitty Hawk dunes to witness the event. Wilbur piloted the first motorized flight in history, which lasted only twelve seconds but was enough to change human destiny forever.

118-119 Arles, France, 1908. The Marquis d'Ecquevilly attempting to fly his elliptically shaped multiplane, with a wooden structure and a metal propeller.

119 Top The multiplane invented in 1907 by John William Roshon, in Pennsylvania. It was a wooden structure of wood and metal tubes with two rows, each comprised of seven levels. It never flew.

COMMUNICATIONS

Camera

Gramophone   Telegraph

Radio

Cinema Television

Telephone

Cell and Smartphone

For thousands of years, humans expressed themselves with their bodies, gesticulating, emitting increasingly articulate sounds, and drawing graffiti in caves. Subsequently, oral communication—it was the bards who passed on the cornerstones of European culture like the Homeric texts—became written, immortalized on papyrus and in illuminated manuscripts. In the fifteenth century, the advent of Gutenberg's printing with movable type made it possible to reproduce texts on a large scale. The earliest books were printed, first the Bible and then all the others: the vernaculars spread, and with them knowledge, which became more accessible, passing literally from hand to hand.

However, in the twentieth century, everything changed rapidly. Various inventions arrived on the scene. There was the gramophone, with which one could listen to Enrico Caruso's powerful voice even before he became a legend in America; the camera, the faithful companion of fearless photo reporters or brazen *paparazzi* looking for scoops; the telegraph, which enabled America and Europe to communicate instantaneously, thanks to long cables crossing the Atlantic; and Morse Code, composed of dashes and dots (information arrived in real time, which had never happened before!). Around 1920, radio invaded homes with music, news, and advertising, establishing itself as the first mass communication medium. The cinema had been widespread for years, and was continuously evolving after the first shaky images of the Lumière brothers. Later, television came into every living room, and marked the daily rhythms of the family. That "talking box" oriented consumption, formed thought and spread ideas, and transformed language. In the meantime, the telephone enabled people—simply by lifting the receiver and dialing a number—to reach anyone, wherever they were.

It was a short step to the cell phone, an essential and multifunctional device, on which we spend hours and hours of our day, using many ways to communicate: sending a photo, a voice mail or an email, reading the latest news, listening to music, and watching a film or a television series. Today, anyone who doesn't download the most popular app, who doesn't take a selfie to post on social media, or who just sends text messages, is regarded as an alien who has just landed on Earth, or rather as a dinosaur who survived a meteorite... We must always try to remember, however, that the *sine qua non* of communication is having someone else who listens.

Camera

View from the Window at Le Gras required eight hours' exposure. It was taken by Joseph-Nicéphore Niepce and is considered the first photograph in history; it reproduces the view from the window of his home in Saint-Loup-de-Varennes, France. Since 2007 the house has been a museum.

When light captures the moment

We can draw with light. The first to have this intuition was Joseph-Nicéphore Niepce, who recorded the image of roofs and buildings from his French home in 1826. At the beginning it seemed to be magic, that a reproduction of reality could materialize in that wooden box with the shutter, a pair of lenses, and a light-sensitive silvered copper plate. But the art of freezing time on paper had a long development and many originators. Not until Louis Daguerre, in 1839, would the image become sharper and capable of being printed. With the first daguerreotype, in the Old and in the New Continent, painters laid aside their brushes and became photographers, above all of portraits. However they were non-reproducible single prints. For a leap into the era of duplication we needed the British William Fox Talbot, a few years afterwards, with the idea of the matrix: a negative on paper soaked with silver nitrate and sodium solution from which to print thousands and thousands of positives, all identical.

It was George Eastman who in 1888 rendered obsolete the plates and elephantine machinery with a roll of flexible film, already packaged in a camera. "You press the button, we do the rest" was the slogan accompanying this little prodigy. It began in the United States and was called Kodak, which does not mean anything but has a good sound and is easy to remember.

Once it was simpler and more practical to use, the optics continued to be perfected, and Americans and Germans were in the forefront for professional cameras.

Already at the beginning of the century, the Graflex, the single lens reflex by Zeiss, hung from the neck of photo reporters and in the trenches, to portray the valorous soldiers of the First World War; in their uniform pocket they preserved the gaze of the woman they loved in a photo. In 1925, the curtain went up on the Leica I: weighing 12.3 ounces, 35mm film—like that used for motion pictures—with a 50mm fixed-focal lens. Kodak and Minolta added color in the mid '30s, and a few years later, the first Polaroids printed a snapshot seconds after clicking the shutter.

124-125 Boulevard du Temple, Paris, 1838. Reducing the exposure time to about seven minutes, Louis Daguerre immortalized, perhaps without intending to do so, a shoeshine and his customer. It was the first photograph of human beings.

125 Camera for taking daguerreotypes, built by G. Knight & Company (London), ca. 1860. Top: a portrait of Louis Daguerre at 61.

Camera

Spectacular pinups in bikinis and the horror of Vietnam, family portraits in an interior and the beautiful stars of Hollywood: images of reality, but also fiction, airbrushed figures and montages. In the second half of the twentieth century, the Japanese companies Nikon and Canon competed for the market and, shortly before the end of the Millennium, another Japanese company, Sony, announced the beginning of the digital era. Today everyone is a photographer, but photography remains an art, banal and magical today as well, being an almost automatic technological process. It is always light that calls the shots.

126 William Fox Talbot busy with negatives and photographic paper at his establishment in Baker Street, Reading, together with his partner Nicolaas Henneman and young apprentices. From 1844 on, the studio also offered services to the public, among which printing from negatives and portraits.

127 Trafalgar Square, London, 1844. Talbot's photo portrays Nelson's Column under construction, during an interval in construction; the calotype process invented by Talbot himself enabled him to reproduce an infinite number of copies from the same negative. On the right, the Mousetrap (ca. 1835), made of wood and brass, as nicknamed by Talbot's wife.

128 Portrait of Henri Cartier-Bresson with a Leica, in 1956. The French photographer, co-founder of the Magnum photographic agency, was the pioneer of photojournalism, whose first camera was the celebrated Leica I.

129 The RB Auto Graflex (left) with which in 1922 the Egyptologist Howard Carter documented his discoveries in the tomb of Tutankhamun in Egypt. A Rolleiflex (right) was among the first reflex cameras with a double lens; it was compact and precise and made entirely of metal.

130   The camera lands on the Moon. This Hasselblad model—the Lunar Surface SWC—dates from 1968. It was designed and modified specifically for the Apollo missions, starting from the 70mm film, which was achieved by Hasselblad itself. It was capable of taking sequences of photos at short intervals; the special ultrathin film had 200 exposures and the new Zeiss lenses involved minimal distortion.

131   Alan Bean, an astronaut on the Apollo 12 mission, during his moonwalk in 1971; a Hasselblad camera mounted on his spacesuit recorded his movements while he gathered fragments of lunar soil. There were many cameras available to the crew, and some of these were abandoned to make space for lunar "souvenirs". The astronaut who took the photo, Charles "Pete" Conrad, is reflected on the

Camera

Enrico Caruso, the great Italian tenor, examines a Victor record. He is leaning on a gramophone placed inside a record cabinet. With his *romanze*, Caruso was the true prophet in song of the Gramophone Company.

Gramophone

Caruso's voice enters the home

The white terrier was called Nipper, and he listened curiously to the voice recorded on the roll. It issued from the horn of a phonograph. According to the legend, that voice reminded Nipper of the voice of his master, who had died a few months before. The master was the brother of the French painter Francis James Barraud, who took up his brushes and captured the scene. Two years later, the painting happened to be seen by Eldridge Johnson, the engineer and founding partner of Victor Talking Machine. He offered Barraud a hundred pounds to replace the phonograph with a gramophone: the deal was done and the image became the unforgettable logo of Victor Talking Machine: His Master's Voice. It was 1901, in the middle of a war for the recording and reproduction of sound.

On one side was the greatest collector of patents ever, Thomas Alva Edison—the creator of the first technological research center, Menlo Park, New Jersey, through which passed half the technical revolutions that would change life in the twentieth century. On the other, a stubborn German immigrant to America, Emile Berliner, whose technical right-hand man was Eldridge Johnson: together, they later founded the laboratory in Camden, also in New Jersey.

On one side, the phonograph cylinders made of wax and tinfoil, in which Edison would believe until the end. On the other, Berliner's disks, the first of crystal glass, ink and lampblack, then of zinc and wax, and finally of shellac on metal plates, which would make their debut shortly before the end of the century.

The official history relates that wax cylinders were the first to reproduce a human voice in 1887. In reality, the first to record a voice was the Frenchman Édouard-Léon Scott de Martinville: he recorded *Au clair de la Lune* on a machine called a phonautographe, 20 years before Edison. But Menlo Park at that time was a black hole where the ideas and progress of a thousand inventors ended up, for technicians and lawyers to transform them into patents in the name of Thomas Alva Edison.

134 One of the first gramophones conceived and produced by Emile Berliner. The serial number was 45048, the year 1890, the make Berliner Gramophone. It was an uncomplicated machine, with all the mechanisms visible, but already very efficient.

134-135 A Victor gramophone, with the metal horn mounted directly on the tone arm, and the white dog listening intently. The pay-off is "His Master's Voice". It is one of the most famous and successful advertisements in history.

His Master's Voice

In 1902 Enrico Caruso recorded 10 *romanze* at the Albergo di Milano, a short walk from the Scala for the astronomical figure, once again, of 100 pounds, for the Gramophone Company of London (a partner of Victor). In 1904 in New York, he would record *Vesti la giubba* (On with the motley) from Leoncavallo's *Pagliacci*: he would be the first singer to sell a million copies of a record. It would make a fortune for Victor.

Records had won the battle, and Edison would surrender only after a long struggle in which he had invested a considerable amount in his cylinders. Columbia Graphophone Record and, in France, Pathé were founded. Then came RCA, and the other majors of the music industry until Sony, and beyond. Next came vinyl records, the microgrooves, better known as "LPs". They were unbeatable.

Gramophones carried to the darkest corners of the world the voices and sounds of artists and orchestras, the speeches of great political leaders, the prayers of various representatives of God on the Earth. Since then, no one has really been alone.

136 The standard phonograph model, the first with the Edison brand, was sold from 1899. It was inexpensive, efficient, and used 2 inches diameter rolls. It was one of the most successful.

137 In the Edison phonograph advertisement, the woman is holding a Gold series cylinder in her left hand. This model of home phonograph and the Gold cylinders were presented as having extremely faithful reproduction.

Telegraph

How Samuel Morse opened up long-distance transmissions

"What hath God wrought": with these words the modern era of communication began. The date was May 24th, 1844, when this Bible verse was used in the first official telegraph transmission, from Washington to Baltimore.

Before then, Napoleon had used the so-called "Chappe telegraph" to transmit encrypted information during military operations; however, the system was based on visual messages, which were sent by towers with a long arm and received at a distance with the help of a telescope.

It was not until the '30s of the eighteenth century, with discoveries in electromagnetism and the transmission of electric impulses, that the American Samuel Morse perfected an electric system of communication associated with an on-off code, which took its name from him. In this way, it was possible to transmit about 500 words an hour. Other scientists were proceeding with similar experiments across the ocean. In 1837, the British Charles Wheatstone and William Cooke put into operation an alternative telegraph system which ran along railroad lines. But the simplicity of Morse's system ensured its international success, beating all competitors.

In only a few years, the usefulness of the telegraph was obvious to everyone: it transmitted official and private communication, as well as press agency news, but only by land. In 1845, for the first time, a cable encased in rubber was laid under the sea and finally, in 1866, after some failed attempts, the telegraph crossed the Atlantic, linking Europe and the United States in real time. Weeks were necessary for thousands of miles of cables to be laid on the ocean bed by an ocean liner.

The first electric telegraph was patented by Wheatstone and Cooke in 1837. It was based on a system of five needles and six keys, and allowed the transmission of messages with 20 letters of the alphabet without the use of a code. It was the first telegraph to operate commercially.

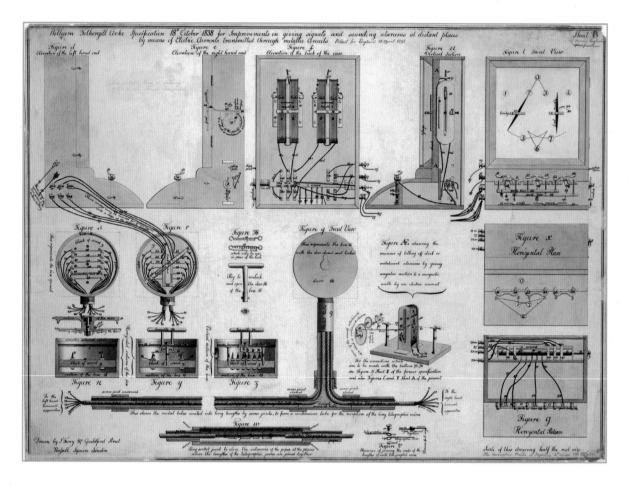

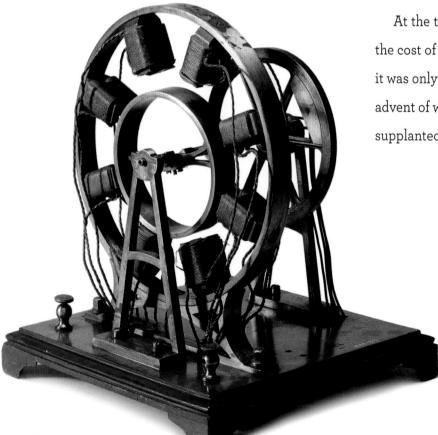

At the time, about 3,000 messages were sent every day at the cost of 5 dollars per word. It seemed like a revolution, but it was only the beginning. The Morse Code even survived the advent of wireless telegraphy, when the first radio broadcasts supplanted the telegraph. But this is another story.

140 Top The figure illustrates William Cooke's studies to improve the electric telegraph.

140 Bottom One of the first prototypes of the "wheel" telegraph developed by Charles Wheatstone.

141 The two-needle telegraph that in 1845 brought about the arrest of a famous murderer, the first case in history, thanks to the use of technology.

In a photo from 1942, the cadets of Tuskegee Airmen, in Alabama, learn to send and receive messages by wireless telegraph. Knowledge of Morse Code became an integral part of military training, especially in the Air Force.

A lesson in the military aviation school in Neuruppin, north of Berlin. Although voice communication by radio was already very widespread, wireless telegraphy was still used during the Second World War to transmit encrypted messages.

Telegraph

Radio

Independent and rebellious, this means of communication survived until the third millennium

Dot dot dot, dash dash dash, dot dot dot. SOS. It was the early hours of April 15th, 1912 when this message in Morse Code was sent from the radio room of the Titanic, toward the open sea. While the sinking of the Titanic indelibly marked the collective imagination, at the same time it turned the spotlight on the use of "wireless telegraphy", which became an essential accessory on great ships and other craft. Only a few years before, in 1901, Guglielmo Marconi had succeeded in transmitting a communication across the Atlantic without the aid of cables. A simple "s", which for the world meant *Straordinario* ("extraordinary") progress.

In 1895, the self-taught twenty-year-old Guglielmo Marconi described his invention in a letter to the Ministry of Posts and Telegraphs and asked for financing for his research. It must have seemed like science fiction to his contemporaries. The request was labeled with the phrase *alla Longara* (meaning "insane"). Italy was not ready to welcome such a revolution, and it would be Britain that recognized his patent for "Improvements in telegraphy and related apparatuses". Radiotelegraphy became first "wireless" communication system. Although history gives the credit to Marconi, many claim to have invented it, first of all Nikola Tesla, who competed virtually in parallel with him on the other side of the Atlantic.

Guglielmo Marconi busy with a radiotelegraphic device. Regarding the invention of the radio, Charles Proteus Steinmetz remarked: "Before Marconi presented his invention to the world, no one would ever have believed that he could achieve it, while afterward made claims to having invented it before he did."

The record for sending a voice signal over a mile, however, goes to the almost unknown Canadian Reginald Fessenden on December 23rd, 1900. In reality, the invention of the radio was not the conquest of a solitary genius, but the result of teamwork, a "remote" endeavor by some of the most brightest scientists at the time, who independently followed the same intuition: by exploiting "Hertz waves" (i.e. electromagnetic, or "radio" waves) it was possible to transmit messages in the ether, without wires. War revealed the potential of radio to the world, and states lavished funds on the development of the most advanced technologies, which after the war were applied in civilian life. The first regular radio broadcasts began. It had never happened before that a message could reach, immediately, a universal and potentially infinite audience. It was the birth of mass communications, for the first time truly democratic and accessible to all.

Radio came into the home, with the exact time, the news, jazz, and advertising. Our way of life was changing and radio was one of the protagonists of the change. For modern middle-class families it assumed the role that fireside stories had

146   Scientist, and prolific and eccentric inventor, Tesla collected approximately 300 patents during his life. It seems that Marconi himself was aware of his studies when he performed his experiments on radio waves. Both achieved identical results nearly simultaneously: in 1900 Tesla was awarded an American patent.

147   On Christmas Eve, 1906, Reginald Fessenden broadcast the first radio program in history: a Handel aria, passages from the Gospel, and a personal violin performance were broadcast along the Massachusetts coast over a radius of 15,5 miles. The radio was born.

Radio

148 Radio was an instrument of propaganda for the great dictatorships of the '20s and '30s, but the radio also carried President Roosevelt's reassuring "fireside chats" (we see him in this photo with his mother Sara and his wife Eleanor, in 1936), which he addressed to the nation during the Second World War. The US President used radio to communicate with the people: by expressing his opinions and conveying security during those difficult years, he was able to come into people's homes and gain the trust of the Americans.

149 In 1931, even the Catholic Church moved with the times: Guglielmo Marconi was given the task of building a dedicated radio station. Vatican Radio was opened by Pius XI on February 12th, with a message in Latin that reached across the ocean. In this photograph of April the same year, Marconi (the first from the left) is in the radio control room with the technical team.

Radio

played in the past. The radio is capable of transformation in appearance and content, and it has been on the crest of the wave for decades. Contained in elegant and voluminous radio sets, then in even smaller and more modern ones, it became portable, found a place in cars, and then defined the world and imagination of young people who went wild with rock 'n' roll. In the hands of great designers it was clothed in seductive shapes and intense colors, and then became rebellious and clandestine to express the desire for change and political protest of the sixties generation. In the '80s, it acquired a cassette recorder and was carried on the shoulder.

It felt threatened by the advent of television and then Internet, but it stayed pugnacious, and where necessary it reinvented itself, becoming a luxury, super technological or vintage product. Finally, just as many foresaw its demise, in an umpteenth metamorphosis, it abandoned its old shell to return to its immaterial nature, finding new life on the Web. A hundred years after its appearance, it has not ceased to surprise us.

Cinema

The Lumière brothers set images in motion

On December 28th, 1895, *La Sortie de l'usine Lumière a Lyon* by the Lumière brothers opened a series of 10 short films shown to a paying audience in the Salon Indien du Grand Café in Paris: in the film, workers and clerks crowd outside the gate of a factory. Cinema was officially born. But it was perhaps another film by the Lumières, *L'Arrivée d'un train en gare de La Ciotat*, screened a week later, that marked the revolution of the new technique, the recording of movement. When the steam locomotive came toward the cine-camera, the audience tried to dodge it: some people closed their eyes, some cried out, some got up and fled. The scene has been reconstructed in various films on cinema, and is perhaps a bit romanticized, but it conveys very well the impact of the new "toy".

It was a big step from the static photographic image to filmed sequences. To give the human eye the sense of movement, it was necessary to record and project the images at the speed of at least 16 frames per second. At the end of the nineteenth century, many were working on this, including Thomas Edison, but the Lumières succeeded first, thanks to the invention of the Cinématographe machine, which was used both for filming and

The two brothers, Auguste and Louis Lumière, in their laboratory in Lyon, in a photo from the '30s. From a very early age they had begun to be interested in photography and film at their father Antoine's portrait studio in Besançon.

Color lithograph (1896) by Henry Brispot, preserved in the Bibliothèque Nationale de France, Paris, publicizing the Lumière brothers' invention, the "Cinématographe".

for projection, and the spooling of a perforated film. The cine-camera and projector worked with a handle, with 35mm celluloid film, which was highly inflammable; the Lumières began to produce it for themselves, but already at the end of the nineteenth century Eastman Kodak dominated the market.

What happened afterwards was a succession of new inventions that stimulated the birth of the film industry. The uttermost symbol became Hollywood, created in America in 1911, but film production grew rapidly in Europe as well. The cinema expanded its narrative ability through the evolution of direction, scenography, camera techniques, and montage. The first films documented reality: wars, political demonstrations, entertainment, nature, daily life; then new genres appeared. The comic genre would give us the masterpieces of Buster Keaton, Charlie Chaplin, Stan Laurel and Oliver Hardy; the historical, epics like *Battleship Potëmkin* (1925); in America, the "Roaring '20s" produced comedies by Cecil B. DeMille, the epic *Ben Hur* (1925), Westerns. The next decade was that of the great stars who filled the tabloids and people's dreams: Rodolfo Valentino, Greta Garbo, Marlene Dietrich.

The advent of sound marked a turning point and the speed increased to 24 frames per second. From the first silent films, with on-screen captions and accompanied by live music from the pianist or even an orchestra, films moved to original music, but still recorded separately, and finally to *The Jazz Singer* by the director Alan Crosland, the first talking film produced by Warner Bros in 1927. For the color revolution, despite the first manual experiments and the

Two frames from the first films projected by the Lumière brothers: *La Sortie de l'usine Lumière* (left) from December, 1895 and *L'Arrivée d'un train en gare de La Ciotat* (right) from January 6th, 1896.

154 Standing on the left, we see the American director and producer Cecil B. DeMille on the set of one of his first films, ca. 1920. De Mille became famous for a series of Biblical films, especially *The Ten Commandments* (1923).

155 Greta Lovisa Gustafsson, whose stage name was Greta Garbo, at the peak of her success, on the set of the film *The Painted Veil* (1934), taken from Somerset Maugham's novel and directed by Richard Boleslawski. The actress, of Swedish origin, retired from acting at the age of 36.

various procedures attempted later, it was necessary to wait for the Technicolor system in the '30s. It was first seen in the Walt Disney animated film *Flowers and Trees* (1932), and it would remain linked to popular films like *Gone with the Wind* (1939) or *Heaven Can Wait* (1943) until it was replaced by the more economical Eastmancolor in 1950, which is still used today.

Today the Seventh Art is going through a crisis caused by new means and techniques of communication, but research and experimentation are always opening new frontiers: 3D films, computer animation and immersive cinema, where the effects created in the auditorium add a fourth sensory dimension. Cine-cameras and projectors have become digital, and traditional film is disappearing, although many great directors, like Spielberg, Loach and Tarantino, still defend it. There are fewer movie theaters and films are seen on the Internet, but cinema continues to produce unique testimonies and timeless masterpieces.

Television

The US inventor Philo Farnsworth presents one of his first models of electronic television. He registered his first patent in 1927, and during his life he would register 300 more, including one for a small system of nuclear fusion.

The magic box that dominated twentieth-century broadcasting

U.S. Patent 1.773.980 of January 7th, 1927. These are the number and deposition date of the American patent destined to revolutionize the information and entertainment industries of the twentieth century: the television system. The owner of the patent was Philo Farnsworth, twenty-one years old at the time, who had already been developing his design of electronic television for six years. An engineering *enfant prodige*, the son of a Mormon couple from Utah, in his life he would accumulate more than 300 patents, almost all of them concerning transmission systems.

The invention of television was a long time in the making and underwent innumerable modifications: it can certainly not be attributed to the work of a single genius. Farnsworth's system, which used a cathode ray tube (invented in its turn by the German physicist Karl Ferdinand Braun in 1897) proved to be more effective and durable. Just two years had passed since the Scottish engineer John Logie Baird had deposited a patent for an electromechanical television system, which was not destined to achieve commercial success. However, it is remembered for a "first": the first television image sent over a distance (from one room to another of his laboratory). It was October 2nd, 1925 when the distorted head of a ventriloquist's doll appeared on the receiving screen; immediately afterward, Baird called his errand boy, William Edward Taynton, to be a guinea pig. He was thus immortalized as the first television "star".

Thus the inventions really mounted up. In fact, Baird's television was in turn based on research by the German optician Paul Gottlieb Nipkow who, while still a student, in 1884 built a perforated disk capable of analyzing and reproducing images; and another German physicist, Arthur Korn, who by using selenium cells had succeeded in 1906 in transmitting a photograph over a 1,100 mile distance. In 1939, it was again John Logie Baird who made a fundamental contribution to the electronic television, when he invented the Telechrome system, the precursor of color television, which spread in the '40s. At the turn of the millennium, technological evolution accelerated swiftly: the miniaturization of electronics, plasma screens, LCD technology, digital television, and satellite broadcasting; all these innovations no longer bore the name of individual inventors, but were the product of research laboratories controlled by electronics multinationals.

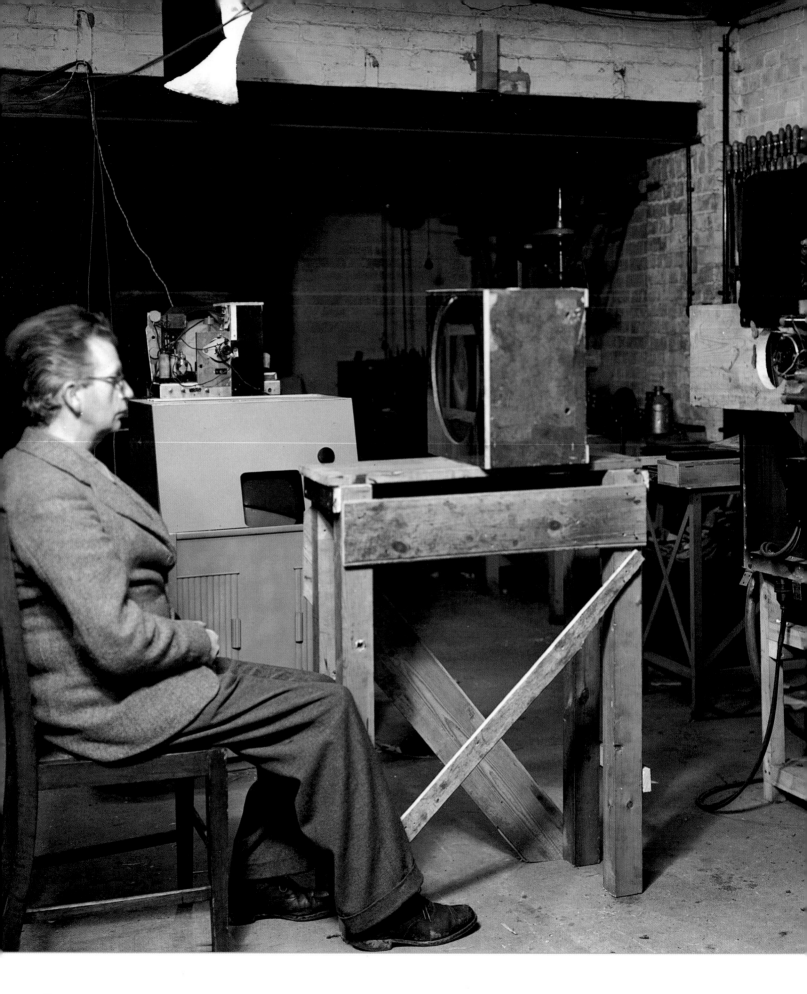

Television has been very popular right from the beginning. The first public broadcasts took place in the United States in 1928: the programs showed moving faces and toys. The first French television company began in 1929. In Britain, the BBC began broadcasting on a daily basis in 1932, then Moscow in 1938 and EIAR in Italy (the ancestor of RAI) in 1939. The nation states, aware of the potential (also political) of the new medium, very soon created state broadcasting companies. In Germany (the home of Telefunken, the first large European television producer) the Nazi regime used television for propaganda: the Berlin Olympics were the first world event seen on TV.

158-159 and 159 The Scottish engineer John Logie Baird in a 1941 photograph; Baird was the first to transmit an image over a distance with an electro-magnetic system, and in color. In the photograph (right) we see one of the first images transmitted by Baird.

160 An American housewife in front of the television set in the mid '50s: it was an image symbolizing the postwar economic boom. The first television companies in the United States were NBC, which began broadcasting in 1946, and ABC and CBS, starting from 1948.

161 An historic broadcast by ABC, which on July 21st, 1969 covered the event of the year: the Apollo 11 moon landing. In the image we see the astronauts Neil Armstrong and Buzz Aldrin planting the American flag on the surface of the Moon.

Television

Then the war slowed down the spread of television sets, but immediately afterward there was a exponential increase in sales: while in Britain in 1947 there were 15,000 family televisions, already in 1951 there were 1,400,000, and this would rise to 15 million in 1968. In the next year, the moon landing would attract the largest audience ever: over 500 million viewers. At the end of the millennium, the television was the most common object in homes throughout the world: an electrical appliance capable of influencing the language, consumption, political orientations, and morals of its users. However, its spread is now decreasing: new electronic devices have begun to take the place of the classical television, above all portable computers and smartphones. The new frontier is interactivity.

One of the first rudimentary telephones built by Alexander Graham Bell. It is 1915, the year in which the first intercontinental telephone call was made, between the United States and Europe: from Arlington, Virginia, to Paris.

Telephone

Alexander Bell and Antonio Meucci: the voice comes down the wire

Sometimes, we still use the word "telephone" for the object we use to communicate, to take photographs, to listen to music, and to surf the Internet. But it no longer resembles what we had at home and "simply" allowed us to talk to people even thousands of miles away: "E.T. Phone Home", asks the extra-terrestrial in Spielberg's film, to contact his fellows and return home.

We must go back to 1876 to trace its origins. Alexander Graham Bell, of Scottish origin, the Professor of Voice Psychology and Diction at Boston University, obtained a patent for "an apparatus for transmitting the voice and other sounds by means of electric waves", in short, a telephone. By means of a microphone, the voice (or rather the acoustic vibrations it produces) was transformed into variations of electric current that reached the receiver of another apparatus, where it became voice again. The Morse telegraph, with its dashes and dots, was already obsolete after only 30 years.

But what about that Italian exile with a mania for science, the friend of Garibaldi, who lived on Staten Island, New York? Of course. Antonio Meucci—that was his name—who in 1871 had deposited, at the patent office in Washington, a caveat (a provisional application, because he could not afford any more) for an invention called "telettrofono", in short a telephone. Thus the official patent bore the name of Bell: English mother tongue, with marked entrepreneurial gifts and important connections, he quickly became a millionaire. Meucci was only left with regret and rage; for the rest of his life he was consumed with suing Bell, in vain.

The telephone revolutionized the work of businessmen, stockbrokers, and post office clerks. Soon families owned one, too: relatives and distant friends were closer because they no longer needed to write letters that would take weeks to reach their destination.

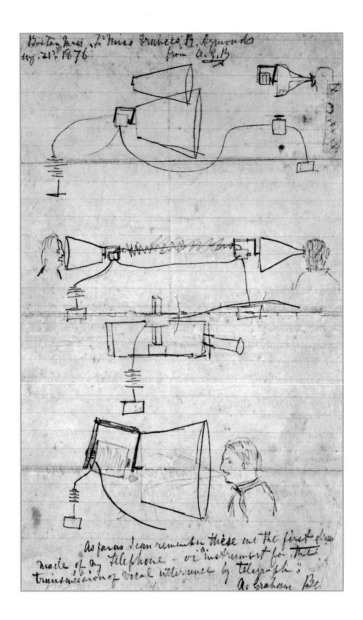

Routines and social relationships immediately adapted: endless telephone calls between friends (especially girls), secretive ones to girlfriends (*I Just Called to Say I Love You*, sang Stevie Wonder), and the call to the lawyer in crime films. And that's not all: switchboard operators, telephone tokens, and telephone booths, the remains of a world that no longer exists. On the other hand, it's increasingly difficult to find a telephone that's only. . . a telephone.

164   The working of the telephone simply explained, in a sketch by Alexander Graham Bell (1876), the year in which he obtained the patent. It is preserved in the Library of Congress in Washington.

165   October 18th, 1892. Businessmen are present at the launch of the 900-mile telephone line from New York to Chicago. It was performed by Alexander Graham Bell himself, the holder of the telephone patent.

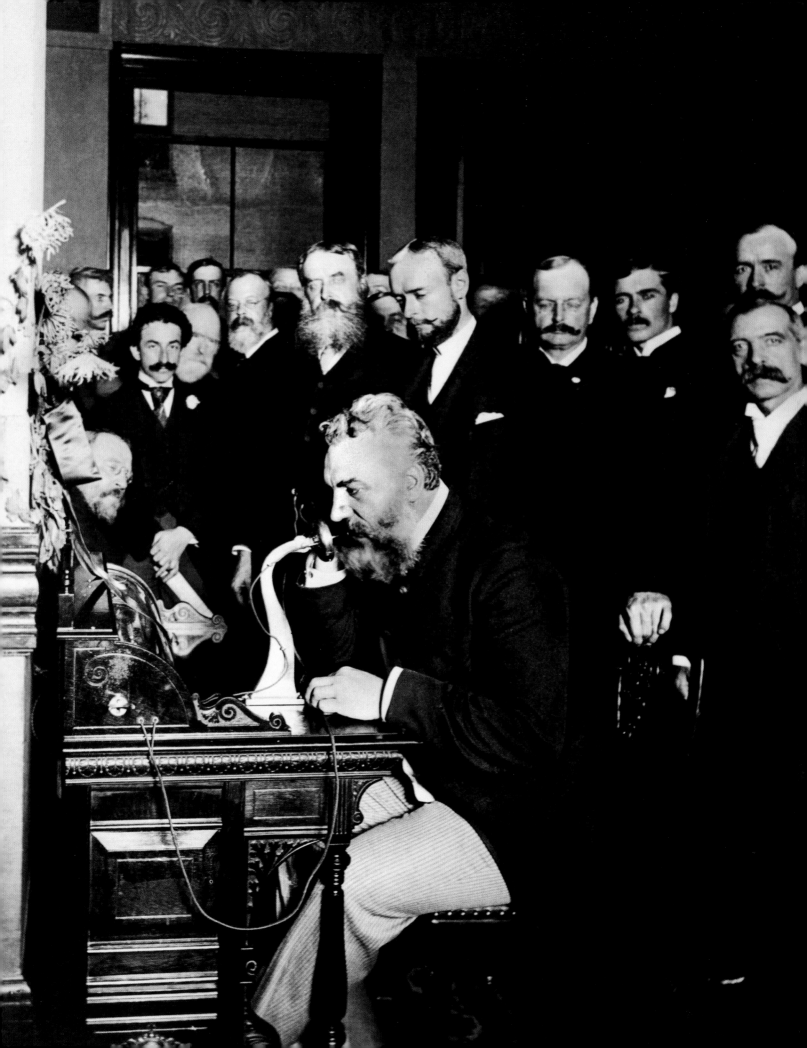

Before the cell phone was invented, there were various solutions to the need to telephone on the move. In 1935 ca., a boy at Naples station has a telephone around his neck to give travelers the chance to make calls upon arrival or departure.

Cell and Smartphone

The telephone leaves the home for the pocket

It is the first thing we grab in the morning when we wake up and the last thing we make sure we have before leaving our home. And if we are out and we realize that we have left it at home, we are deeply worried: the cell phone has thus become an integral part of our life, and we should not be surprised if, in the not too distant future, it were implanted under our skin as a microchip. At least we would be sure we had not forgotten it.

Already at the end of the nineteenth century, Lars Magnus Ericsson, the founder of the company of the same name, felt the need to have a portable device with which to phone at any moment: as he was afraid of his car breaking down, he took a phone to connect to the nearest telephone pole in the street.

The first real call from a cell phone took place on April 3rd, 1973 from a busy street in Manhattan, amid yellow taxis and skyscrapers. Martin Cooper, who was then an engineer at Motorola, placed to his ear a device which one could have mistaken, in weight and dimensions, for a brick. With a touch of pride he dialed the number of Joel S. Engel at AT&T, his direct competitor.

After 10 years of experimentation in limbo, the cell phone arrived in the stores: the Motorola DynaTAC 8000X was far from pocket size and took up most of the space in the cases of wealthy entrepreneurs who could afford to buy it. However, the real turning point was the change from analog to digital: with the GSM standard, there were also SMS (text) messages. This was in 1992, and this cell phone service had an incredible success; with its 160 characters it got everyone used to being more concise.

From the end of the '90s, while public telephone booths inexorably disappeared, the cell phone was the protagonist of an incredible surge: reduced dimensions and lower costs led to its success around the world, followed by new standards in communication, the first hesitant connections to the Internet, new services like games, the camera and video calls.

A revolution, which was soon surpassed in 2007 when Steve Jobs launched the iPhone in a spectacular presentation. The first model of smartphone was a device that did everything with captivating design and, for the first time, had a touchscreen. It was practically a pocket computer. Everything changed. New words poured into our daily vocabulary: app, emoji, selfie, hashtag, follower, Whatsapp. It changed our everyday life and our social interactions. Everything was transformed in the name of freedom to communicate. Or perhaps of a new form of slavery.

168 and 169 Bottom The first cellular phone, the Motorola DynaTAC 8000X, made by Martin Cooper in 1973; it was 9 inches long and weighed 2.5 pounds; the battery took 10 hours to charge and lasted 30 minutes. There was no comparison with the 1996 pocket model, the Motorola StarTAC (bottom right), the first real mass-market telephone.

169 Above In 2007 Steve Jobs launched the iPhone, presenting it as a revolutionary cell phone that combined internet and iPod functions with a touchscreen.

MEDICINE

Aspirin

X Rays   Vaccines

Penicillin DNA

The Pill

Pacemaker

From the last decades of the nineteenth century, Western medicine contributed considerably to the rise in population that took place in the twentieth century as it drastically reduced infant mortality, improved the quality of life and increased longevity. What we relate in this chapter represents the tip of the iceberg of this great story, which has seen science gradually overcoming superstitions, social taboos, and religious dogmas that had survived for centuries.

In this sense, the early pioneers of the previous centuries played a fundamental role, after the long, dark period of the Middle Ages, and investigated the human body and its functioning with open minds.

Thus, in the sixteenth century, Andreas van Vesel, thanks to the dissection of cadavers, carefully described the internal organs. A century later, William Harvey identified the directions of blood circulation through the body. The era of the microscope began with lenses built at the end of the seventeenth century by Antonie van Leeuwenhoek. In the first half of the nineteenth century, Ignaz Semmelweis identified in bacteria the origin of puerperal fever, against the resistance of the traditional Academy. More or less in the same period, anesthesia was developed.

It is at this point that our story begins. Vaccines were the first effective weapons with which to fight the terrible epidemics that decimated the population until the first decades of the twentieth century; with antibiotics, the modern pharmaceutical industry was created, and aspirin, the first synthetic drug, is still one of its symbols. In the meantime, diagnostic techniques have made use of electromagnetic radiation to have increasingly precise images of the interior of the human body, and technology is the source of medical devices, like the pacemaker, that assist in the treatment of serious diseases. The contraceptive pill was one of the great social turning points of the twentieth century; organ transplants also gave new hope to the seriously ill, and studies on genetic material allowed scientists to focus on diseases that were previously incurable. This is not without the ethical problems that increasingly accompany cutting-edge research. Scientists already imagine the possibility of defeating death through the modification of genes or cloning, and there are already those investing in the dream of rebirth after hibernation.

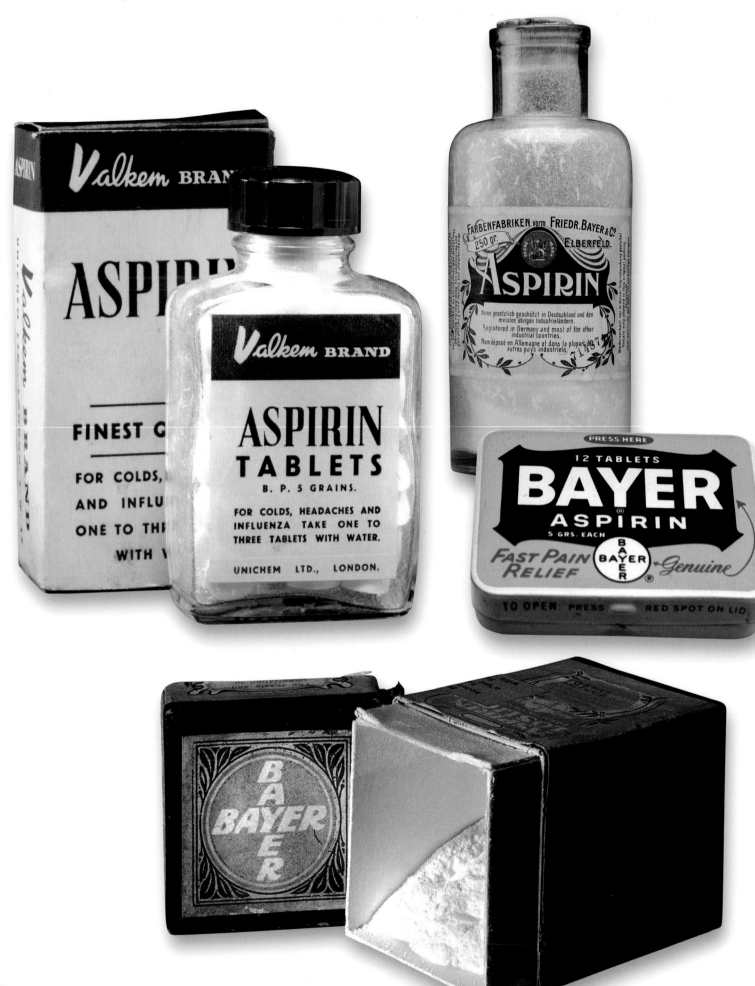

Aspirin

The symbol of the universal remedy

A white pill, to swallow or dissolve in water. For more than a century it has been the medication *par excellence* for pains, inflammation, colds and influenza. Aspirin boasts several records: the first synthetic pharmaceutical product, the first to undergo clinical trials, the most sold in the world (100 billion pills sold per year), and the first to reach the Moon with the astronauts of Apollo 11. Its origins are the foundations of the modern pharmaceutical industry, and its name is synonymous with Bayer, which still remains its main producer. But not everywhere in the world, because in some countries the German company had to give up its patent following the liberalization imposed by the Allies after the First World War.

While the active substance has been used since ancient times, until recently we did not know the precise mechanism of its action. *Salix alba* (white willow) and meadowsweet, or *Spiraea*—from which the name "aspirin" derives—are two plants that are widespread throughout Europe and the Mediterranean basin; the ancient Egyptians and the Greeks knew their therapeutic properties. The willow fell into disuse in the Middle Ages, when its branches were valuable for baskets and it was forbidden to harvest it. But when Napoleon, at the beginning of the nineteenth century, stopped the importation of quinine from the Americas, the *Salix alba* was rediscovered. The extracts also worked against fevers, but it was still not known how to isolate the active substance. Salicylic acid was isolated in 1828 by Buchner, but its most interesting derivative, acetylsalicylic acid, was synthesized only in 1853 by the French chemist Charles Frédéric Gerhardt. However, we were still far from production of a well-tolerated drug. The first acetylsalicylic acid that could be used for therapeutic purposes was called ASA, which we owe to Felix Haffmann, a researcher at Bayer. The patent was registered in 1899.

Alone or associated with other active substances, synthetic acetylsalicylic acid was marketed in various forms around the world as an anti-inflammatory, analgesic and antipyretic. The appearance of other drugs like Paracetamol and Ibuprofen in the '70s and '80s caused a decrease in sales, which have recently risen again thanks to Aspirin's protective qualities for the heart.

The Man who doubted if HOWARDS' ASPIRIN was the BEST

See the name Howards on every Tablet.

174 Advertising poster from the '30s for the pharmaceutical company Howards of Ilford, north-west of London. After the First World War, Bayer lost its exclusive right to the name in many countries.

175 French advertisement from the '30s for aspirin produced by the chemical company Usines du Rhône, which had already marketed the drug in 1908 under the trade name "Rhodine", before the trade name "Aspirin" was derestricted.

Aspirin

The scientist who clarified the process was John Vane, who received the Nobel Prize in 1982 for his research on inflammatory processes. Thus, another important property of acetylsalicylic acid was discovered, that of inhibitor of the complex system leading to blood coagulation. While in excessive doses this characteristic may create problems or bleeding of the gastric mucus, in the right doses it can prevent the formation of clots. Scientists are already thinking about the use of aspirin to reduce the risk of heart attacks and strokes, and as a protection against some forms of tumor. As the British researcher Collier predicted, it is a molecule whose "future is more promising than its past."

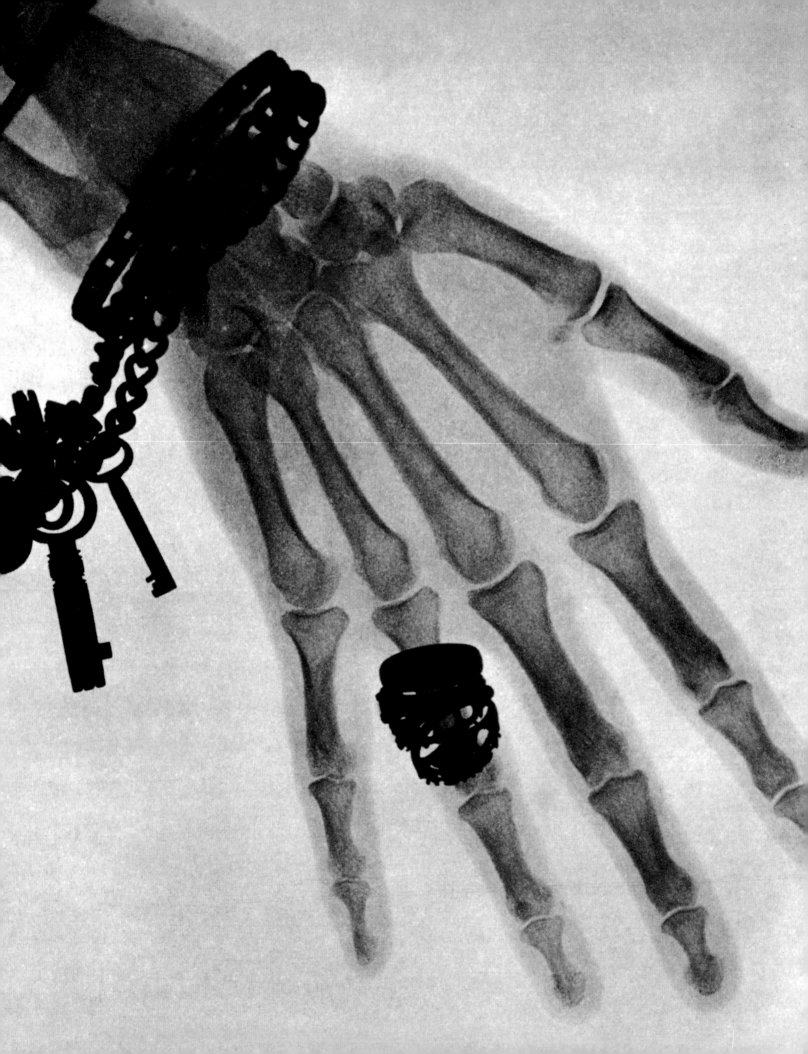

X Rays

Wilhelm Röntgen and his chance discovery

X for unknown. Although these electromagnetic radiations are familiar to us, we still call them by the name given in 1895 by the German physicist Wilhelm Röntgen, who discovered them by chance and left us as testimony the first X-ray in history: his wife's hand, on which the wedding band is clearly visible.

Röntgen was performing experiments with a cathode ray tube, a glass cylinder in which a vacuum was created and which contained two metal plates, the cathode and the anode. When these were subjected to a high electric potential difference, the flow of electrons leaving the cathode generated radiation when it reached the anode. Röntgen noted that this invisible radiation was able to pass through many materials and leave an impression on a photographic plate; but not all materials—not metals, in particular lead, or rigid structures like bones. Thus radiology was born, and it immediately found its application in medicine.

To be able to see inside a human body without having to open it surgically represented a formidable step forward for diagnostics; any fracture could be studied from the exterior and often treated without using more invasive techniques. Very soon, however, it was discovered that this radiation with a very short wavelength—from 0.01 to 10 nanometers, little more than the devastating gamma rays—was capable of interfering with organic molecules and in massive doses was very harmful: many researchers paid the price until protective measures with adequate materials were adopted. Röntgen was honored with the Nobel Prize for physics in 1901, but refused to patent his discovery and died in poverty.

One of the first X-ray images, taken in Britain around 1896, only a few months after the announcement of the discovery by Wilhelm Conrad Röntgen. We can clearly also see a ring, two bracelets and a key chain; radiation, in fact, does not pass though bones or metals.

Since then, while diagnostics has taken huge steps, such as revealing in detail structures like the heart and lungs—which normally would not be able to block X-rays—and culminating in modern tomography (TC), this radiation has even found an application in astrophysics, with telescopes able to observe the universe through a window different from that of visible light and to gather a large amount of information on the structure and mechanisms which govern its evolution.

178 Wilhelm Conrad Röntgen, a professor at the University of Würzburg, in his home-laboratory where the evening of November 8th, 1895, discovered electromagnetic radiation, for which he coined the name "X-rays".

179 Two mobile units during the First World War. The first is a German tent for surgical operations, in which we can see the equipment for performing X-rays with the Röntgen method. The second is a radiography unit employed by the US Army.

Vaccines

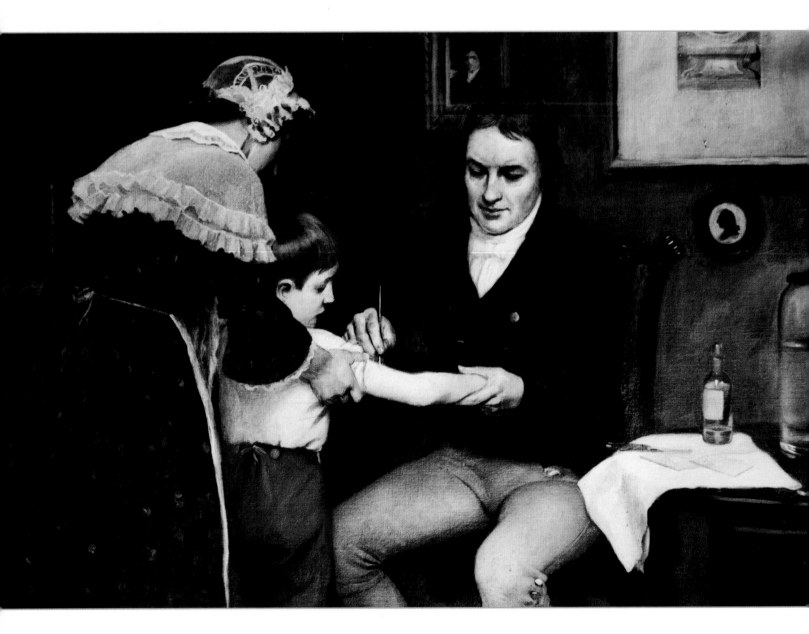

In this painting by Ernest Board, Edward Jenner tries, for the first time, the anti-smallpox vaccine on a human being. It is May, 1796: when Jenner presented his results, the scientific community was initially hostile to his discovery.

Edward Jenner, Louis Pasteur, and the struggle against widespread epidemics

The virus, an invisible and multiform enemy, which was responsible for various epidemics which periodically decimated the world population, was not identified until the end of the nineteenth century. But already a century prior, scientists had intuited its effects and sought to fight it with its own weapons. The first vaccine was perfected by the British doctor, Edward Jenner, against smallpox, which at the time was the worst of all scourges afflicting Europe. Observing that the milkers who came into contact with cowpox did not contract the disease in a serious form, Jenner thought of injecting material obtained from the bovine pustules into humans, so as to set off the same process of immunization produced by every organism against infection. In fact, the body reacts by producing particular proteins called antibodies, which are able to communicate to the immune system the extraneous elements to eliminate. It is a race against time, because viruses and bacteria reproduce very rapidly but, when the organism prevails, it is cured and preserves within itself some "memory cells", sentinels able to protect it from subsequent contagion.

Acting on the same principle, at the end of the nineteenth century Louis Pasteur's vaccine vanquished rabies; in the mid '50s, the vaccine Sabin blocked the spread of poliomyelitis; and in 1979 the World Health Organization was able to declare that smallpox had definitively disappeared. Vaccination is still the best system to contain epidemics, but it must be widespread; it is called the "herd effect": only when a group reaches a certain percentage of immune individuals is the entire group protected. Not all infectious diseases have been defeated, however, because viruses and bacteria change continuously, producing variants resistant to vaccination; this is true of the HIV virus, against which so far there has been little success.

Moreover, in many countries in the world, conditions are lacking to combat epidemics. Thus, in the same way that Jenner had to defeat the prejudices of his time regarding a vaccine of animal origin, every era has its battles and the most effective weapons, apart from vaccines, are the simplest ones: the struggle against poverty, hunger, ignorance, for the right to clean water and decent sanitary conditions.

182-183 Vaccination against rabies at the Institut Pasteur, founded in Paris in 1887 by Louis Pasteur, who was also the first director.

183 Louis Pasteur, pictured here, also perfected the process of pasteurization to eliminate pathogenic micro-organisms from some foodstuffs.

New York, 1962. Campaign by the Department of Health for the distribution of the Sabin vaccine against poliomyelitis. It had been approved in the same year, but it was perfected between 1953 and 1955 by the Polish American doctor Albert Bruce Sabin.

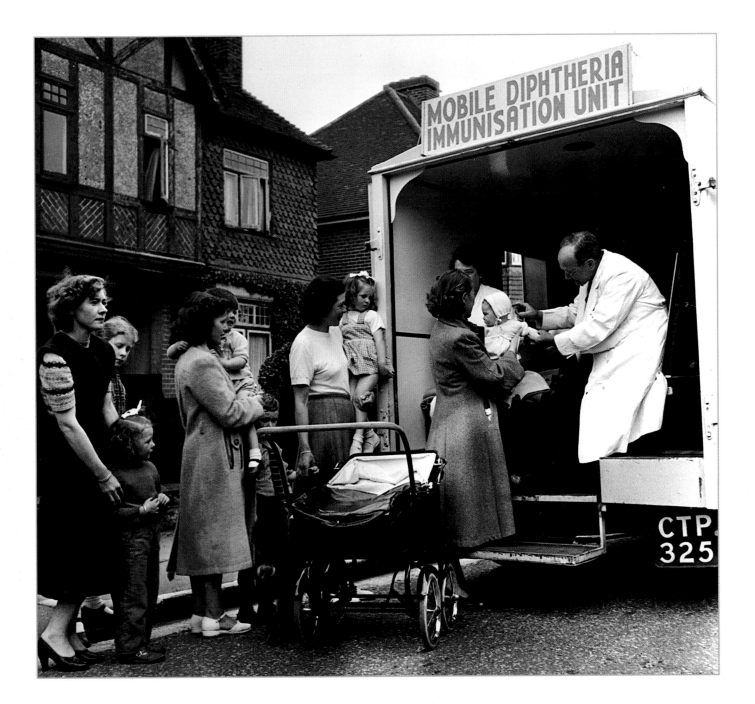

Portsmouth, 1951. The general immunization campaign by mobile units of the National Health Service drastically reduced the number of cases of diphtheria in the United Kingdom. The doctor Emil von Behring, who discovered the vaccine against diphtheria and tetanus, was awarded the first Nobel Prize for medicine in 1901.

Vaccines

Penicillin

How Alexander Fleming changed the history of medicine

While the Great Powers fought on the battlefields of the Second World War, another war was being waged in the research laboratories on both sides of the Atlantic. Even if there had been no catastrophic epidemic like the influenza of 1918-20, the spreading among civilians of pneumonia, rheumatic fevers, typhoid, and cholera owing to lack of hygiene and food—as well as the condition of soldiers at the front, who were subject to deep wounds and amputations—required more powerful antibacterial drugs than the traditional remedies, which until then had been quinine and arsenic.

While Germany concentrated on sulfonamides, which had been discovered in the '30s by a physician named Gerhard Domagk, the Allies' "secret weapon" was a mold whose effectiveness had been known for more than 10 years, but which was difficult to produce in the necessary quantities. We owe the discovery and the name "penicillin" to the British doctor and biologist Alexander Fleming, who in 1928 had observed the inhibition of the growth of the Staphylococcus bacterium in cultures accidentally contaminated by a fungus of the genus *Penicillium*. Those who isolated and purified the active substance were the Oxford researchers, Howard Florey and Ernst Chain, in 1940. They demonstrated its effectiveness by numerous studies *in vivo*, and for this they were awarded the Nobel Prize, together with Fleming.

The research was begun in Britain, but it would need American financing and an unprecedented effort in order to launch, around 1945, mass production of penicillin, and continue in the following years with the formulation of other antibiotics. All of this would radically transform the pharmaceutical industry. Years later, these drugs that have saved millions of lives have also revealed their limits: the need, now demonstrated, to increase doses to maintain therapeutic effectiveness, coupled with their sometimes excessive use can, in fact, produce alterations in the organism's bacterial balance and the appearance of super-resistant bacterial strains.

Alexander Fleming intent on studying mold cultures in his laboratory in the Inoculation Department of Saint Mary's Hospital in London, today the Wright-Fleming Institute, named in honor of Fleming and the microbiologist Almroth Wright, who led its initial immuno-biological research projects.

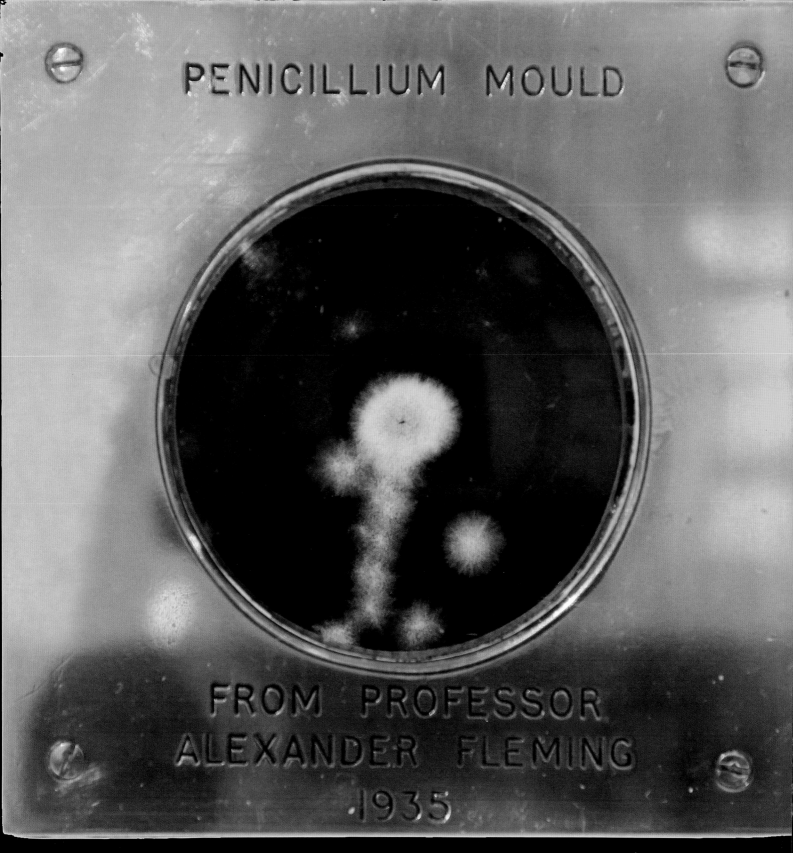

PENICILLIUM MOULD

FROM PROFESSOR
ALEXANDER FLEMING
1935

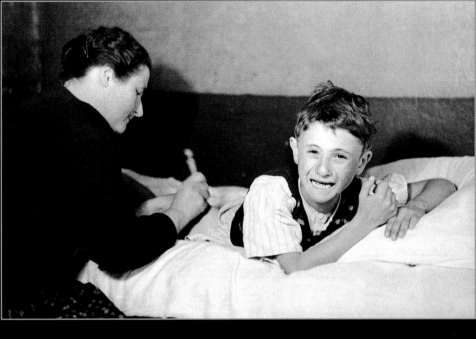

188   A sample of penicillin exhibited on November 4th, 2009 at the Science Museum in London, on the occasion of an exhibition celebrating 100 years of science. *Penicillium* is a relatively common genus of fungus present in many foodstuffs; penicillin is produced from the species *Penicillium chrysogenum*.

189   The photo published by the British magazine *Picture Post* on December 18th ,1948 shows a young boy being treated with an injection of penicillin by his mother, in the French village of Mont-près-Chambord.

DNA

Investigation into the origins of life

"It's a history book, a narrative of the journey of our species through time. It's a shop manual, with an incredibly detailed blueprint for building every human cell. And it's a transformative textbook of medicine." In 2001, the geneticist Francis Collins, director of the Human Genome Project, thus described the mapping of the human genome. Two years later the work was complete, with the identification of about 22,300 human genes, a number markedly lower than the one hypothesized at the beginning.

The project began in 1990, involving the USA, the United Kingdom, Canada and New Zealand. However, right from the beginning of the century, geneticists had investigated the mechanisms of hereditary transmission, starting with the American Thomas H. Morgan, who was awarded the Nobel Prize for identifying chromosomes as the depositors of genetic information. The turning point was in 1953, when scientists began to understand the structure of DNA and decipher its language. Two British institutions worked on this assiduously, Cambridge and King's College. On one hand, the two researchers Crick and Watson, on the other Maurice Wilkins and the young Rosalind Franklin. Franklin was the first to "see" the structure of the double helix and to photograph it with X-rays; Crick and Watson constructed a rudimentary model of it, and in 1962

Analyses of DNA conducted on individuals from all the continents have indicated a 99.9% genetic identity of the human species, despite physical differences. 94% of the remainder (0.1%) concerns variations within the same population, and only 6% concerns variations between individuals from different populations.

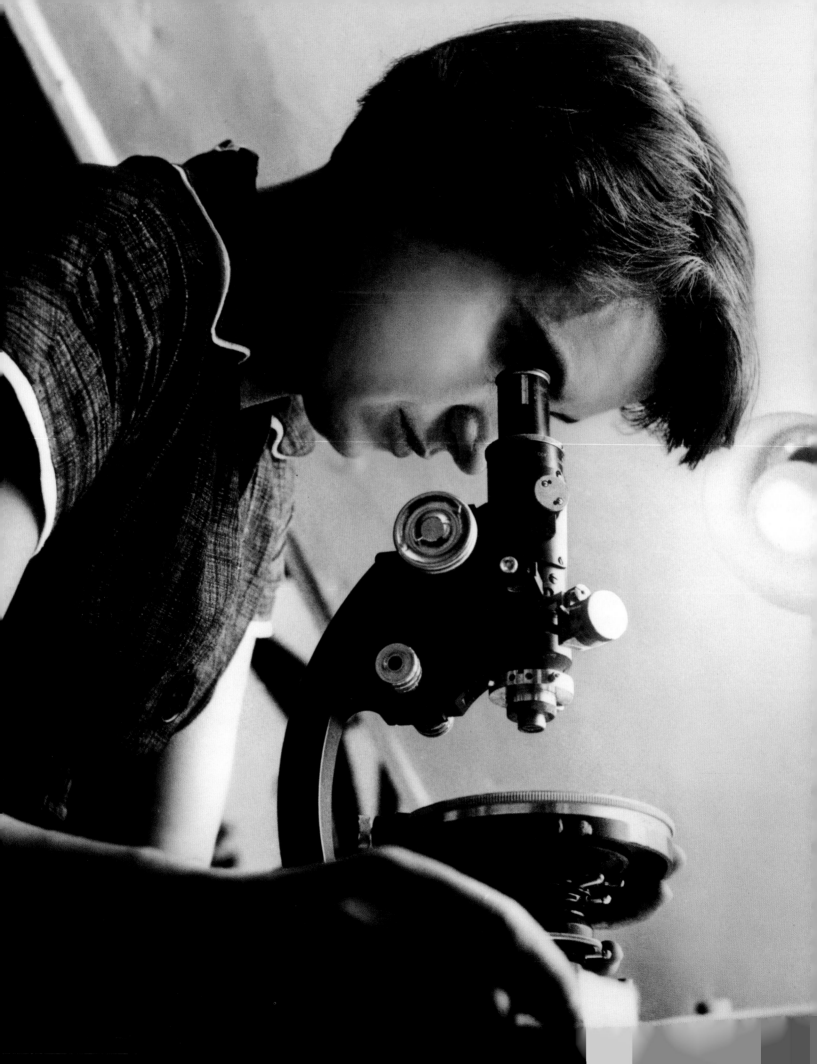

received the Nobel Prize with Wilkins, while the unfortunate Rosalind Franklin, who had died in the meantime, was not even nominated. Watson, who in his career was often in conflict with the scientific world for his intolerant ideas, would be criticized for this as well.

In fact, the work published with Crick remains fundamental in showing how nitrogenous bases form two extremely long helixes that intertwine and at the moment of cellular reproduction are compacted and transform into chromosomes. The sequences in which the nitrogenous bases are arranged constitute the genes. Determining the order of the sequences was another success achieved in the '70s with the contribution of various scientists, which earned two Nobel Prizes for the British Frederick Sanger.

192   The British chemist Rosalind Elsie Franklin, an expert in X-ray crystallography, at work at King's College, London. The images of DNA that she was able to obtain between 1951 and 1952 are the basis upon which Watson and Crick determined the double-helix structure.

193   James Watson (left) was only 23 and Francis Crick (right) 35 when they began to work together at the Cavendish Laboratory in Cambridge. Less than two years later, they announced the discovery of DNA, presenting a model constructed with cardboard and wire.

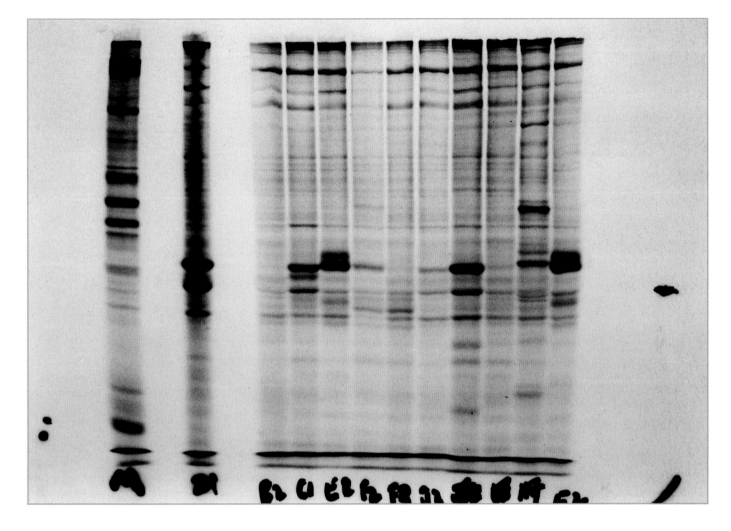

Today, the completion of human genome mapping opens up new paths for molecular biology and medicine, which can determine an individual's genetic risk, identify genetic variants responsible for various pathologies, and in some cases intervene by modifying sequences of DNA. The analysis of genomes studies microbic communities in their environment, and in forensic medicine the DNA test enables investigators to work back to the culprit from a small biological sample. Evolutionary studies establish affinities and correlations among different species and even analyze extinct groups, making discoveries like the recent ones by the paleogeneticist Svante Pääbo on *Homo neanderthalensis*, who proved very similar to *Homo sapiens*.

The studies by the American doctor Harold Elliot Varmus on the genetics of the retrovirus, by means of electrophoresis on the gel separating the molecules, show the mutations of a DNA infected by the ASV (avian sarcoma virus, the cancer of connective tissue), 1979.

The Pill

The woman chooses whether to be a mother

For thousands of years, the biological mechanisms underlying conception remained a mystery, and birth control methods, handed down from mother to daughter, combined superstition, magic, intuitions, and pseudoscientific misconceptions. Today, it seems atrocious to insert into the vagina crocodile dung or stones, to eat lizard tails cooked in mercury, and even to drink water used to wash dead bodies; yet in some cases these methods made some sense, because many of these substances acted on the acidity of the vaginal area and on sperm activity.

There is an endless list of herbs used since the Middle Ages to prevent pregnancy and produce an abortion, often with dire effects. It was necessary to wait for the twentieth century, after centuries of false myths, abortions and infanticides, for the first partially effective contraceptives to appear, at least in the Western world. In the '30s, contraceptives, intrauterine devices, and the Ogino-Knaus method existed; scientific studies had reached the point where everything necessary to create a hormonal contraceptive was in place. And yet, no one was interested in taking the research forward, because it conflicted with laws prohibiting birth control and the hostility of the Catholic Church. Until in 1951 the biologist Gregory Pincus met the activist Margaret Sanger: he was an expert in sexual hormones and worked for the pharmaceutical company Searle, while she had the motivation and the economic resources to encourage research on a technique that could completely limit fertility.

Margaret Sanger is seated in the center; around her are women from her Brooklyn clinic in the mid '20s. She was a pioneer of reproductive rights, the first to introduce the term "birth control" and awareness of the question into the English-speaking world.

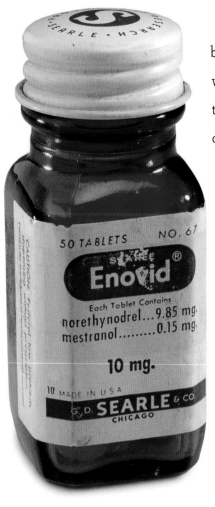

Six years later, Enovid began to be marketed in the United States. It was a tiny pill distributed "for the treatment of menstrual disorders"; only later it would be used as a contraceptive, and for some time it was allowed only for married women. It also arrived in Europe, with the name Anovlar. For everyone, it was destined to become simply "The Pill".

Despite its small size, its social impact has been powerful and deep: it has enabled women to separate their sexual lives from their reproductive, family, and affective lives—as is the case for men—with unprecedented assurance.

Its invention, despite doubts surrounding its experimentation and the initial disregard for side effects, was a conquest that marked an era. It arrived just in time for 1968 and the revolution in social mores, of which it was bound to become a symbol.

198 and 198-199 In 1916 Sanger was put on trial for having opened a birth-control clinic. In the photo (right) she is immortalized coming out of the Court of Special Sessions in Brooklyn. Her campaign would influence the debate on the legalization of contraceptives in the United States: Enovid (top), the first pill, would be marketed by Searle in 1957.

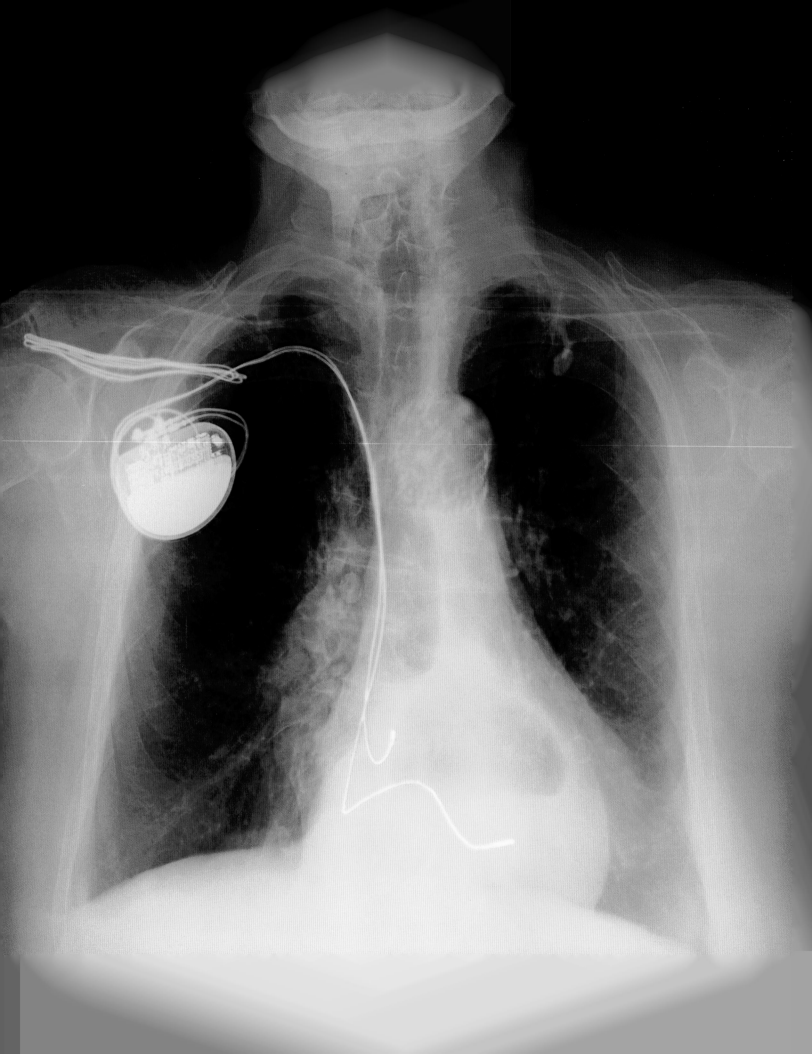

Pacemaker

William Greatbatch, the engineer who "set the pace" for the human heart

Some wires and an electric generator in polished metal, weighing less than one ounce, which does no less than "set the pace" of the human heart: the pacemaker has quickly found its place among the inventions marking the history of the twentieth century, so much so that the National Society of Professional Engineers has declared it one of the greatest social contributions of engineering in the last 50 years. In fact, it was conceived by an American engineer named William Greatbatch. Born in 1919, he had been a radio operator in the U.S. Navy during the Second World War before earning a degree in electro-technical engineering. From the articles published upon his death in 2011, he seems to represent the classic inventor: well read, curious, passionate, endowed with a spirit of observation, a fertile imagination and exceptional intuition. He must have been so, if during his life he managed to register more than 325 patents! The studies to which he dedicated himself, from renewable energy sources to HIV treatment, demonstrate his attention to social problems, to environmental challenges, and to the quality of life. Since 1960, when it was first implanted in humans, it is calculated that the pacemaker has contributed to saving millions of lives the world over.

As with many discoveries, this happened by chance, or rather through error. Greatbatch was working on an instrument to record cardiac rhythm and reveal possible anomalies. Rather than record heartbeats, it was the improper application of a component that made the device produce similar impulses to the beating of a healthy heart. From this, he took the idea of using the device to "condition" the human heart by means of a series of electrical impulses that stimulated the contraction of atria and ventricles, according to how it was programmed.

With practically non-invasive intervention, 1-3 electro-catheters were inserted into the cardiac cavities and the stimulator was placed in a pectoral "pocket" under the skin. The latest technological developments have produced pacemakers without electro-catheters, increasingly smaller and with ever lighter materials, broadening the range of functions, from prevention and treatment of arrhythmias to remote monitoring: a great step forward for human longevity and quality of life.

X-ray of a patient in whom a pacemaker has been implanted. The patient is affected by a hiatal hernia, a protrusion of the stomach that can affect the heart, producing fibrillation and arrhythmia.

INFORMATION TECHNOLOGY

Computer

Video Games

Internet

It is all around us, everywhere. It is so pervasive and invisible that we no longer even notice its presence. If one day, by some strange aligning of the stars, it were to disappear, we could not even imagine the disastrous consequences. And yet everything that governs our lives today, which seems increasingly complicated, is an extreme simplification: an infinite sequence of 0s and 1s—the binary code—which is performed faster than human thought processes.

The need for an instrument to perform calculations has existed for thousands of years: from the abacus, the first means for making calculations, to mechanical calculators, like those of Pascal and Babbage, centuries passed with new inventions, until we arrived at the visionary mind of Alan Turing. He imagined a universal computer, programmable through algorithms, to perform an infinite number of tasks. From this point, machines became capable of processing information. And a little late, in the '60s, the new science was given the name of "information technology".

The inventions that this section discusses were the product of many small intermediate stages: without microchips, transistors and video terminals, we would not have had personal computers, smartphones or PlayStations. It is a road with an unknown destination, both distant and unpredictable. But even if we cannot foresee the future, we can observe the impact that IT has had on our daily lives: the simplification and acceleration of innumerable tasks, but also alienation in the virtual world of videogames, the endless and universal reach of Internet, the revolution in social relationships on Facebook.

New questions open on the horizon: what future awaits us? Augmented reality, artificial intelligence, quantum computers? With mixed enthusiasm and apprehension, we wonder whether humans will be able to make use of machines to surpass their limits, or they will lose control and become slaves to them, as prophesied in many dystopian films. Despite the fact that people speak of implanting microchips and various devices in the human body, we know that the best performing machine is the one that we already have inside our head. Who more, who less.

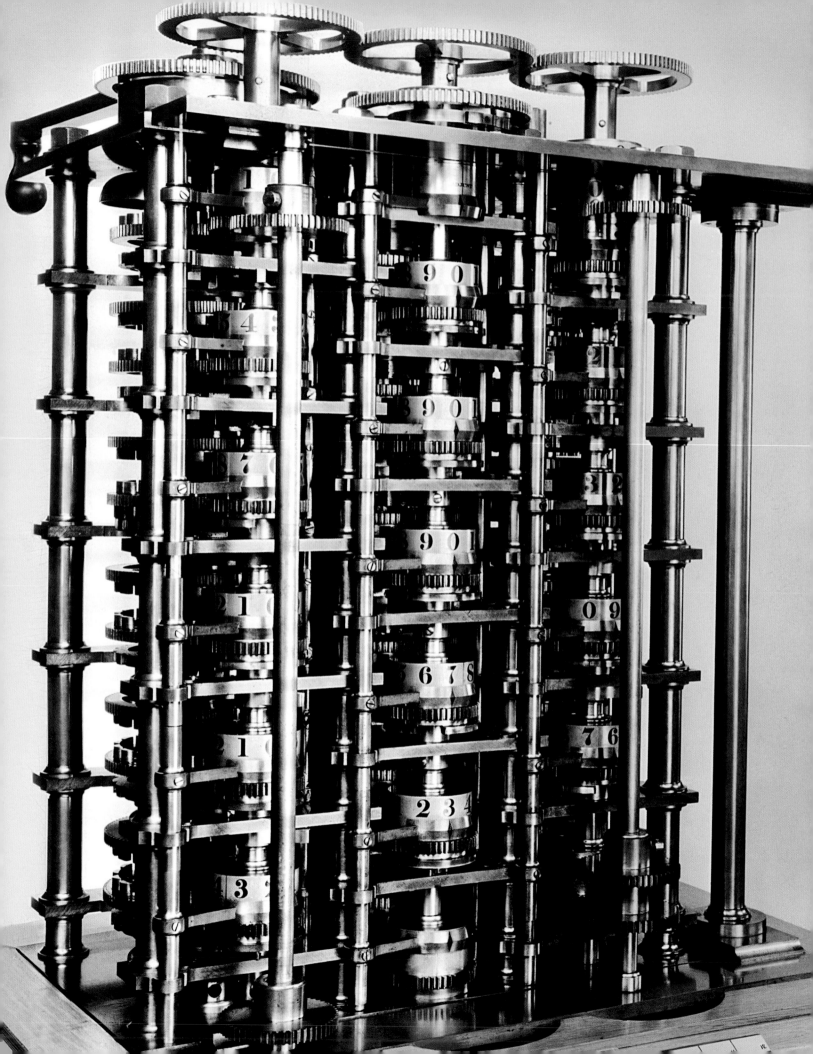

Computer

From Alan Turing's vision to the Bill Gates-Steve Jobs contest

"Can machines think?" Alan Turing asked this question at the beginning of his most famous paper, and over a few decades the question would be debated over and over.

It was the '30s, and the machine had taken the name of "calculator", because it had been conceived to perform calculations; soon it would explode in less than half a century from being a giant made of cables and diodes, as large as an apartment, to being that small, portable, and multifunctional box—the smartphone—which today we all have in our pocket. It enables us to talk with the other side of the world, take photographs, and record music, and it also recognizes our voice; it takes a source, simplifies it to the nth power in a sequence of 0s and 1s (the binary system), processes it, and gives us a complex result, which would otherwise take a human an enormous amount of time to achieve, because electronic circuits function a million times faster than biochemical ones.

But to modify a well-known proverb, the modern computer was not built in a day: if we wished to trace its family tree, we would see a dense crown with many branches. Terms like Z1, ABC, Mark I, Colossus, and ENIAC could tell the story of many human beings and inventions. Piece after piece, each with its small contribution: the venerable ancestors, like Pascal with the first automatic mechanical calculator, or Babbage with the "the analytical engine" which solved polynomials.

And the white-haired grandparents, like the one that in 1937 was placed in the Berlin salon of Konrad Zuse: an enormous electromechanical calculator that had no memory and read instructions from perforated cards.

Then, as for every great invention, one light shone more brightly than the others: visionary that he was, Alan Turing no longer thought of a simple calculator, but of a universal and programmable calculator, which by means of algorithms could carry out an infinite number of tasks. A vision, in fact, that was still afloat in the world of ideas.

A component of the difference engine designed by British mathematician Charles Babbage. It represents about 1/7 of the complete engine. It was conceived for automatic calculation, and was able to perform additions, subtractions, divisions, and multiplications up to six figures. Today it is preserved in the Science Museum in London.

It was John Von Neumann who grasped it and made it concrete: he was the architect of the solid foundations upon which our computers still stand today. For the first time, with his EDVAC a general-use machine took shape, one that no longer limited itself to doing calculations, and had a structure comprising a central processor, memory units and buses to link all the parts.

But in that era, the machines were as large as closets and occupied entire rooms: then, in the '50s, the first transistor arrived, which was an effective substitute for the thermionic valve—to get an idea of this, we moved from the dimensions of a light bulb to those of a USB flash drive. Then came the integrated circuit: components were ever smaller and faster. In the meantime, some were thinking of developing a computer language: in 1959, Grace Hopper invented the first programming language, Cobol, which we still use today when we take money from an ATM. And then came the '60s: the mouse, but above all the graphic interface, at last making the computer "personal", ready to take its place in everyone's home.

206 Grace Hopper busy on one of the first models of computer, a machine that reads sequences of instructions from a perforated tape.

207 ENIAC was the first digital computer for general applications. Built between 1943 and 1945, it worked with punched cards, weighed 30 tons and occupied a total surface of 200 square yards.

At this point, the revolution knocked on the door of American garages, where some fervid and tireless minds—above all, two: Steve Jobs and Bill Gates—competed by innovation after innovation: their operating systems hardware, and graphic interfaces shaped the computers of the future. It would be the rivalry between Apple and Microsoft, between Mac OS and MS-DOS, in a constant race for improvement, that would go on all through the '80s and '90s, while from Japan came the first portable computer aimed at the mass market. This was soon matched by another great invention on the threshold of the third millennium: Internet, the network that Tim Berners-Lee would launch in 1990 with that "www" we type every day. And the future? Some imagine it as terrifying, like the world of *Matrix* dominated by machines, or by the psychotic HAL 9000 in *2001: A Space Odyssey*, but we are here now, wondering—how will the competition between humans and "intelligent" machines end?

208 February, 1986. Bill Gates founded the Microsoft Campus in Redmond, Washington: only five years had passed since IBM (at the time an IT giant), commissioned him to develop the PC-DOS operating system. A sensational success, it blazed a trail for the first version of Windows, which was launched in 1985. In March, 1986, Microsoft was quoted on the Stock Exchange and the thirty-one-year-old Bill Gates became a billionaire.

209 1984 at the Moscone Center in San Francisco, Steve Jobs and Steve Wozniak, together with the then Apple CEO John Sculley, introduce the Apple IIc model during one of the first spectacular company product presentations.

Video Games

Today, Nolan Bushnell, the father of *PONG*, is engaged with Anti-AgingGames.com in the production of games capable of stimulating memory and concentration in over-35s.

The infinite universes of virtual reality

Imagine infinite worlds peopled by little yellow disks chased by pixelated ghosts, blocks to fit in frenetically, Italian plumbers, shapely lady explorers, photorealistic landscapes, second lives that dangerously invade real ones. And then imagine worried parents, housewives grasping joysticks, broken televisions, countdowns on the calendar, "game over", "continue", little monsters in augmented reality, apps for smartphones, trade shows, top secret previews, ridiculous bugs, patches, social media, laughter, fear, and emotions. All of these have been, are, and very probably will be videogames. A medium created in the '40s without a commercial objective, which in time has managed, for better or worse, to astound us and to generate discussion, to evolve by influencing and stealing cleverly from other media, rethinking them and itself, to create a new way of looking at the world.

In 1972 Andy Capp's Tavern in Sunnyvale, California welcomed the *PONG* arcade game, an elementary, hypnotic table tennis simulation conceived by the figure who, for many, is the father of the videogame industry, Nolan Bushnell. It was the beginning of a revolution, which *PONG* itself extended a few years later by invading television screens in its home version. Over the years, games have appeared that marked an era. To mention only a few: *Space Invaders, Doom, The Sims, Final Fantasy, Halo, Metal Gear Solid*. Today, *Sony* and *Microsoft* constantly compete for the top games, while *Nintendo* pampers its fans with exclusive and innovative consoles, and unparalleled icons that have rightly passed into pop culture. In the middle of it all, right or wrong decisions that have led to videogames being either cast as a demon, or as educational and medical support for combating dyslexia and problems of movement. At the moment, we possess, though we may be little aware of it, a means of communication that is both powerful and mature.

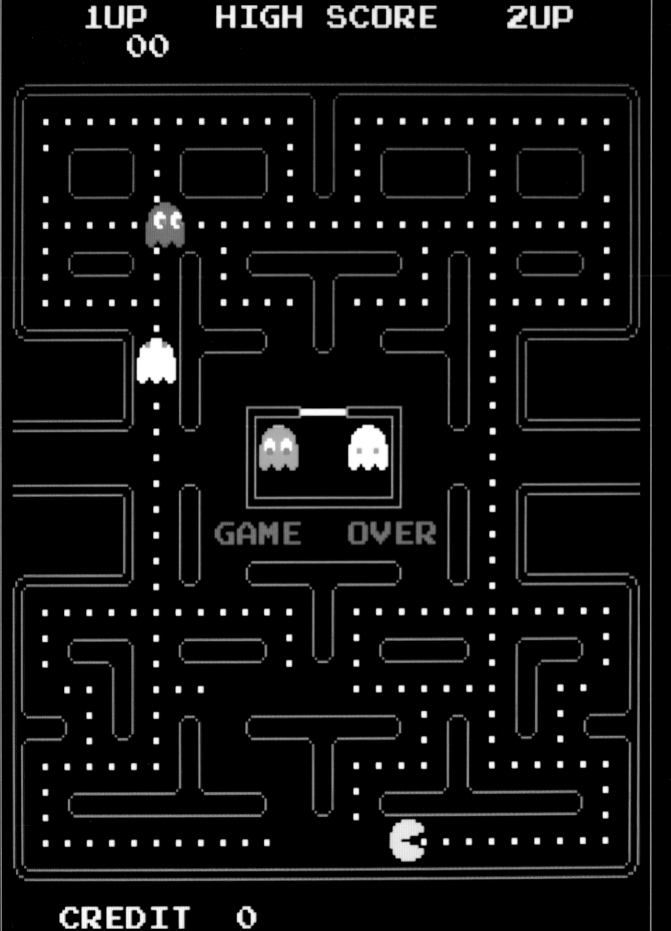

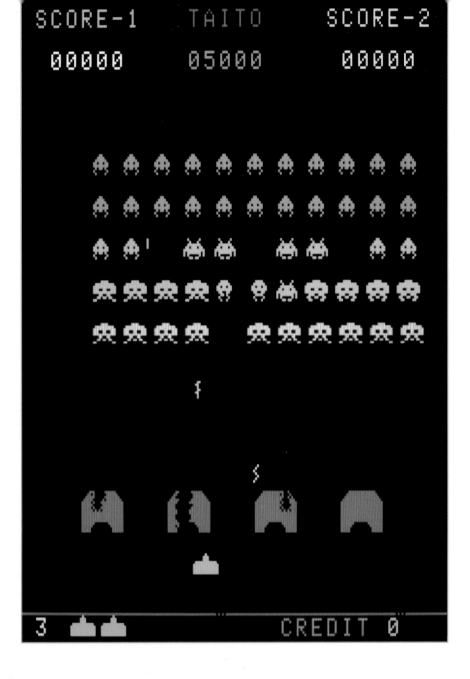

Obama intuited this when in the middle of an election campaign he decided to purchase an extremely personal place in Heaven, and with it, exposure, in the form of virtual billboard advertisements along the digital streets of *Burnout City*. It is a means of communication capable of giving rise to art works exhibited at the MoMA, of creating complex personalities and plots worthy of Hollywood, of creating excitement, and of questioning itself. Certainly, in many ways the world of videogames may not be the best of all possible worlds, but everyone knows that with a few clicks they can return to their own virtual world.

Space Invaders had such an impact that the Japanese had to deal with a lack of 100 Yen coins, which were sucked up by one of the most successful video games in history.

214 With Game Boy, Nintendo got into the pockets of video gamers the world over. And it went even higher: in 1993, it even livened up the space voyage of the Russian astronaut Aleksandr A. Serebrov.

215 Top The Super Nintendo was Nintendo's answer to Sega's Mega Drive. With only 16 bits, the new console gave birth to legendary sagas like *Dragon Quest*, *Zelda*, *Donkey Kong*, and *Final Fantasy*.

215 Center At a price of 595 dollars each, the Commodore 64 holds the world record for the most-sold computer in history, with as many as 17 million units sold. Excellent value for money underlay its success.

215 Bottom At the beginning of the design phase for Super Nintendo, Nintendo turned to Sony to design an external CD reader. However, the two companies could not come to an agreement. Sony continued the development and, in December 1994, presented the first revolutionary PlayStation.

Video Games

Super Mario needs no introduction. The 240 million units sold means that the series of games with the iconic mustachioed Italian-American plumber as its protagonist figures in the Guinness Book of Records as the most successful video game franchise in history.

The Electronic Entertainment Expo in Los Angeles, better known as E3, is one of the most important world videogame fairs and a must-attend event for fans. The most important brands in the field organize conference-shows, in which they give the world a glimpse into the future of the video game industry.

Video Games

Thanks to his innovations in the technological field, in 2016 Tim Berners-Lee was awarded the Turing Prize, one of the most prestigious awards in the field of information technology.

Internet

From research to homes: forever connected

Who could do without Internet today? Work, information, purchases, entertainment, research: it is increasingly indispensable and pervasive. In only a few years, the net has radically changed our daily life and our way of thinking.

It was born in 1969. While all the world was incredulously watching Neil Armstrong set foot on the Moon, an American laboratory was making its first steps on another unexplored terrain: the first computer network.

In that era, right in the middle of the Cold War, the Pentagon was engaged in resolving a fundamental problem for the security of the United States: how to prevent the Russians from interrupting communications among US military bases? The solution was Arpanet, a vast, acentric network of computers with many alternative nodes and connections. Thus, even if there were a breakdown, there would still be a possible route for communication. A few years later came a new system for exchanging messages, a method faster than all the mail carriers in the world. By a quirk of fate, its symbol was a snail. Email was born.

In the meantime, computer networks were springing up everywhere, and it was acknowledged that a common language was necessary to enable them to communicate with one another: a standard protocol was devised, the TCP/IP, so fast and efficient that we still use it today. But the real turning point came in 1991. From a small room at CERN, Geneva, a brilliant young man, Tim Berners-Lee, launched his idea: a great worldwide spider's web, made up of pages, all of them linked to each other and based on a common language, HTML. To access it, one only had to type its initials: www, or World Wide Web.

From that moment, the dizzying, overwhelming, and unpredictable rise of Internet began: starting from research institutes, it came into everyone's home. In step with the technology

220   Facebook was officially launched on February 4th, 2004. Originally designed for students in a few American universities, it soon became a platform open to all. In a few years, it grew staggeringly: in 2017, there were over 2 billion users active at least once a month.

221   On October 11th, 2017, on the occasion of the Oculus Connect 4 event, in San Jose, California, Mark Zuckerberg, the founder and CEO of Facebook, presented Oculus Go, an innovative model of visor aimed at providing widespread access to virtual reality.

Internet

The Internet

(it would be difficult to imagine its spread without Wi-Fi, tablets, and smartphones), every corner of the world, even the most remote, discovered it was online.

After connecting computers, it was time to connect individuals. In 2004, when the daring Harvard student, Mark Zuckerberg, launched Facebook, he did not yet imagine the scope of his creation. Soon, social networks would revolutionize social relationships and change the lives of millions of people. Now we are completely immersed, constantly connected, and reachable everywhere.

We ought to ask whether there is a price to pay, whether our freedom and privacy are in danger. But we are too busy checking whether there is a new notification.

The Author

Gianni Morelli. The author of novels, short stories, essays, educational books, and screenplays, he has worked in research institutes and written books on travel and geography. He has edited various publications for White Star, most recently *1968 A Revolutionary Year in Photographs* (2017, with Carlo Batà) and *The Most Influential People of Our Time* (2017, with Roberto Mottadelli). He is the editor-in-chief at Iceigeo in Milan.

Photo Credits

WHITE STAR PUBLISHERS

WS White Star Publishers® is a registered trademark
property of White Star s.r.l.

© 2018 White Star s.r.l.
Piazzale Luigi Cadorna, 6 - 20123 Milan, Italy
www.whitestar.it

Translation and Editing: Iceigeo, Milan (Jonathan West / James Schwarten, Elena Rossi)

ISBN 978-88-544-1318-4
2 3 4 5 6   22 21 20 19 18

Printed in Italy by Rotolito S.p.A. - Seggiano di Pioltello (Milan)